BRUCE SINCLAIR
with the assistance
of James P. Hull

A Centennial History of The American Society of Mechanical Engineers

1880–1980

Published for
The American Society of Mechanical Engineers by
UNIVERSITY OF TORONTO PRESS
Toronto   Buffalo   London

Canadian Cataloguing in Publication Data

Sinclair, Bruce.
A centennial history of the American Society
of Mechanical Engineers, 1880–1980

Includes index.
ISBN 0-8020-2380-0
1. American Society of Mechanical Engineers –
History. I. American Society of Mechanical
Engineers. II. Title.
TJ1.S56   621'.06'073   C80-094192-6

FOR ALAN DOUGLAS SINCLAIR
1957–1980

I shared my ambitions for this book with my son Alan,
and I knew something of his dreams. But when he began
his punishing struggle with cancer, we became
partners in another kind of hope. A teacher myself,
I learned from him a deeply affecting lesson in courage
and that the most important thing in life is its quality,
not quantity.

Contents

Preface

Americans have believed for a long time that democracy depends on an informed electorate. Yet it is practically impossible for the ordinary citizen to play much of a role in making political decisions about sophisticated technology, whether the questions involve energy policies, environmental considerations, or industrial safety standards. These kinds of problems have usually been left to the 'experts,' professional engineers with specialized knowledge who have also claimed to act in the public interest. Because Americans generally shared with engineers the same ideas about technical advance and industrial progress, in a sense they delegated their authority in such matters to these professionals, who devised codes of ethics as a promise of faithful performance. Of course, it was also in the self-interest of engineers to have that authority and one might say it was as much a case of taking as it was of giving, but in an age of technological innocence the compact seemed fair enough.

Today, the bargain is under serious challenge. A solid phalanx of critics asserts that much industrial output bears scant relation to the public interest, that unsafe products, wanton use of limited resources, and a disdain for the consequences of industry's effluent wastes call for consumer action and governmental intervention. Engineers themselves are certainly conscious of changing attitudes towards technology and of the need for a different approach by the profession to contemporary problems. Donald E. Marlowe, a past president of the American Society of Mechanical Engineers, characterized that awareness in a set of remarks before a joint committee of the United States Congress in 1974. It had long been assumed in America, he said, 'that technological advance was inevitable and that it was always good and that it was limitless. Now we know none of these statements are true.'

Despite such a sensitive appreciation of the changed world in which mechanical engineers find themselves, the situation of their own professional society today reflects the charges technology's critics have so harshly leveled. ASME's tax status as a learned society is under review by the Internal Revenue Service, while in the courts an anti-trust action claims that the organization's boiler code, a technical activity long celebrated for objectivity and held up as a model of public service, has been used for private gain. These events clearly raise questions about the control of specialized knowledge in a democracy and, since the Society has always been fundamentally concerned with generating and diffusing technical information, also about the relation of the public to the work of institutions such as ASME.

The role of history is crucial in any attempt to deal with such problems, another truth Marlowe understood. The Society has been one of the country's most important agencies in the creation of industrial standards, for example, in such vital areas as nuclear power generation, petroleum refining, and machine manufacturing. Thus, its work over the years, as one of a number of collaborating organizations that frame standards, provides a valuable way to study the methods by which the private sector sought to integrate self-interest with larger national concerns. Scientific management was also spawned under ASME's roof and among professional societies it has played a leading role in efforts to apply engineering skills in the solution of a broad array of economic and social problems. Perhaps most significant of all, the Society's history helps to reveal the outlines and consequences of a complex technological information-processing system. It is an article of faith that Americans are an inventive people. But besides machines, they also created a welter of interrelated institutions to translate technical knowledge into industrial practice, and that may have been one of the country's most successful inventions.

My own interest in the relation between technical knowledge and industrial development emerged while writing a study of Philadelphia's Franklin Institute. Founded in 1824, its early history laid the groundwork for many of the activities ASME subsequently pursued. Thus, when members of the Society's Centennial Steering Committee asked me to write its history as part of the anniversary observance, I was glad of the opportunity to continue this line of investigation. I am happy to say that, while ASME commissioned the work, no one ever attempted in any way to influence what historical materials I saw or the conclusions I drew from them. That may be a useful insight for those who automatically think of engineers as only economic men, or simply as the technicians of industrial capitalism.

This book, then, is entirely of my own devising. It necessarily reflects my ideas about institutional history and my conviction that writing about the past is more of an art than a science. For instance, even though the occasion might seem to call for a work of record – a detailed survey of all the different activities the organization has carried on over the past century and a register of the names of all those who have been involved in them – I have instead tried to catch the essential qualities of the Society's existence. My ambition was to identify the well-springs of sentiment and action, as if I were analysing a human life, and it appears to me that, as with personal history, an institution is characterized by enduring themes, most of which are present from the beginning.

Yet no matter how much the actual writing of a book is a matter of individual judgment, it is still a co-operative venture depending upon libraries, colleagues, and family. In that respect, I am pleased to be able to acknowledge the generous assistance of librarians and archivists at Case Western Reserve University, Columbia University, Massachusetts Institute of Technology, Purdue University, Rensselaer Polytechnic Institute, the Franklin D. Roosevelt Library, Stevens Institute of Technology, the University of Toronto, Toronto Public Library, the Library of the United Engineering Center in New York; and also the Secretary of the Engineering Societies of Milwaukee.

Four prominent members of ASME, Dr Richard Folsom, Dr Serge Gratch, Dudley Ott, and Louis Rowley were kind enough to make their own records available to me. I also benefitted from personal conversation with them and with a great many other members whose long involvement in the Society and willingness to speak candidly to an outsider gave me a kind of understanding that documents rarely provide. Dr Henry Black, Dr Donald Marlowe, and Louis Rowley, of the Centennial Steering Committee, proved excellent counselors and their sensitivity in a time of personal trouble has put me much in their debt. ASME staff members, in New York and in the field offices, were always generous with their aid and in a friendly way made my work pleasant and easier. I am especially grateful to Carron Garvin-Donohue and her spirited band of co-workers in Public Relations.

Two old friends, Carroll W. Pursell jr and Michal MacMahon, read and criticized portions of the manuscript and their suggestions improved it. R.I.K. Davidson and John Parry of University of Toronto Press were splendid editors. My colleagues in the Institute for the History and Philosophy of Science and Technology at the University of Toronto also became collaborators in this enterprise, although perhaps not entirely of their own free will. They carried my administrative burdens cheerfully, successfully, and longer than they could reasonably have been asked to do. The contribution of my graduate student James P. Hull, is, I think, accurately stated on the title page. One of my Institute colleagues, Professor Mary P. Winsor, is also my wife and I could not have written the book without her unwavering loyalty.

1

A Sense of History

The American Society of Mechanical Engineers reached the fiftieth anniversary of its founding in 1930. This kind of institutional milestone usually calls for special observance to celebrate an organization's continuity and define its place in history. But besides the conventional impulses for that sort of exercise, the Society's leaders were men with a taste for ceremony and a conviction that their profession deserved a prominent part in the story of human progress. ASME's anniversary festivities thus promised to be rich in the self-congratulation customary on such occasions.

Yet, there could hardly have been a time when mechanical engineers had less to celebrate. Ever since the First World War, science and its applications had come under increasingly critical attack and by 1930 the indictment against 'Machine Civilization' encompassed poison gas, worker alienation, and debased culture. By that time, too, as the 1929 stock market crash settled into economic depression and widespread unemployment, engineers got the blame for the social consequences of over-production, as well as for the ecological and resource implications of their work. America's vaunted industrial capability, the world's most complete expression of modern technology, was on trial and, since they claimed responsibility for it, so were mechanical engineers.

The phrase Machine Civilization came to be widely used in a growing literature during the 1920s and expressed not only the general elements of the debate, but its mood as well.[1] Europeans worried about the Americanization of their continent were most concerned about the aesthetic and cultural

1 For an introduction to the Machine Civilization debate, see Charles A. Beard ed *Whither Mankind* (New York: Longmans, Green 1928), and Stuart Chase *Men and Machines* (New York: Macmillan 1929).

implications of mass production materialism, a position exemplified by Oswald Spengler's book *The Decline of the West*. To other critics, Machine Civilization had come to mean monotonous jobs and bored workers. Instead of freedom from drudgery, machine-tending led to the loss of mental and physical vitality, and threatened a supine work-force more liable to manipulation by demagogues. American engineers, who had always equated technical advance with political freedom and the vigorous culture of the New World, felt challenged by these criticisms, particularly since they tended to come from literary pundits, an intellectual quarter that often looked down on engineering work. But the critique was not simply a matter of aesthetic snobbery. As the Lynds' 1925 sociological analysis of Muncie, Indiana, made clear, mechanization had all but eliminated complex craft skills and the social status which went with them. 'It's "high speed steel" and specialization and Ford cars that's hit the machinist's union,' one older craftsman reported. This observation had special point for an engineering society which had so applauded Frederick Winslow Taylor's careful analysis of cutting tool steels and speeds.[2]

The feeling that things had gone too far, too fast, was also behind the idea of a halt to scientific research and development. That suggestion came from the Bishop of Ripon at the 1927 meeting of the British Association for the Advancement of Science and immediately stirred up a strong reaction from scientists and engineers on both sides of the Atlantic. No public figure seriously argued for the bishop's proposal, however, until the Great Depression suddenly gave new currency to the sense that there was a disturbing gap between technical advance and society's ability to absorb it.[3] To the critics of mechanization, economic calamity seemed the inevitable result of overproduction, magnified by technological unemployment, and much of the responsibility for it was laid at engineering's doorstep. That charge threatened to undermine one of the central justifications of modern industrial technology – the provision of abundance. At a time when millions were out of work, it was much more difficult to persuade people that the chief result of engineering progress was an increased standard of living for the mass of society. On the contrary, Henry Wallace and other New Deal reformers often claimed the social sciences stood a better chance of solving America's problems than the applied sciences, and that argument proved doubly uncom-

2 As quoted in C.W. Thomas ed *Essays in Contemporary Civilization* (New York: Macmillan 1931) 30
3 Carroll W. Pursell has described the controversy in his '"A Savage Struck by Lightning": The Idea of a Research Moratorium' *Lex et Scientia* 10 (1974) 146–61.

fortable for engineers, since it frequently implied the need for an alteration in the country's social and economic system.

But even before the depression gave a new sharpness to the Machine Civilization debate, when planning for the fiftieth anniversary was first underway, the Society's leaders had already felt the sting of criticism keenly and they determined to make the occasion a powerful argument for the importance of engineers and their work. From the beginning, therefore, the celebration was charged with more than rhetorical importance. The nature of the event guaranteed that it would employ certain conventional forms, but the pressure on mechanical engineers to defend themselves also insured that its content would be directly connected to the defense of mechanization. That was clear from the very first report of the Committee on Meetings and Program to the Council on 9 March 1928. The anniversary should be used, it claimed, to demonstrate 'the advancement of industry and civilization as a result of engineering contributions.' To counter the attacks of modern technology's critics, the committee meant to establish the engineer as the primary agent in the upward course of human progress, and its report also imagined the celebration as an opportunity 'for projecting into the future an evaluation of the place of the engineer in industry, in public life and in government.'[4] Subsequent reports to the Council continued to stress the committee's central ambition, to create a program 'on a plane that would not only mark a dramatic pause in the development of the profession but point the way for an ever broadening and deepening influence upon the growth of our civilization.'[5]

The committee proposed to achieve that elevated and expansive kind of celebration through a week-long series of events in April 1930. The main feature was to be a three-day program in Washington, DC, that would include reports by foreign experts on the progress of engineering all over the globe, sessions to visualize the future of engineering, and special fiftieth anniversary awards to the world's outstanding engineers. The committee's objective was to stage an affair so international in flavor and so crowded with distinguished guests that the contributions of engineering to human civilization would be inescapably obvious. It remained only to give the idea a name, although when it came to that, the committee simply selected the most obvious title for the Washington phase of the celebration by calling it 'The Influence of Engineering Upon Civilization.'

The choice of Washington rather than New York, where the Society had its headquarters and traditionally held its major meetings, was another deli-

4   ASME Council Minutes 9 March 1928
5   Ibid 19 Oct. 1929

berate decision. No engineering society was oblivious of the fact that an engineer occupied the White House, and ASME's planners wanted to include President Hoover in their celebration. But the Washington segment of the anniversary commemoration was also the part most explicitly addressed to the Machine Civilization debate, the one most calculated to win public notice, and the committee meant to give it a national setting.

That campaign, however, was only one facet of the fiftieth anniversary planning. It is in the nature of such occasions that they fulfill several functions and the other major part of the celebration, to take place in the New York area, was designed to serve an array of essentially traditional and what might be called internal purposes. For instance, the unveiling at the United Engineering Building in New York of a large bronze plaque, especially created for the fiftieth anniversary, provided a formal beginning for the week-long celebration and newspaper copy suitable to the occasion. The sculpture depicted the forward-looking spirit of engineering in the form of a thoughtful but muscular male figure and it also conveyed an uplifting slogan, 'What is not yet may be,' that its framers hoped would continue to inspire members and visitors to ASME's headquarters long afterwards.

There were other elements of the New York ceremonies that also served sentimental and mythological purposes. All professions have their myths – anecdotal fragments of the past that have less to do with describing historical reality than they do with creating an image of the profession and its practitioners – and the mechanical engineers were no exception. Knowledge of the actual details of the Society's founding, for instance, died with those who had been directly involved, and it is likely each of them saw it differently. But by 1930, a well-rehearsed set of stories about what were seen as the crucial events leading to ASME's establishment had evolved and it was inevitable they would be told again in some form for the anniversary celebrations. The meeting in the *American Machinist*'s editorial offices in February 1880 that first brought together the founders was one such legend. They immediately recognized the need for an association, according to the story, but some dissension among those individualists of strong opinion threatened to destroy the enterprise until, at a dinner thoughtfully provided by the magazine's editor and accompanied by fine wines, good fellowship and co-operation prevailed. The event became an accepted part of the Society's history not so much to explain the organization's creation as to establish for mythological purposes the personal qualities of the founders. It was also just the sort of myth which lent itself to re-creation as a historical pageant and, indeed, the McGraw-Hill Publishing Company presented it that way on the first day of the anniversary celebrations.

But it was not just sentiment that insured the presence of this mythological episode in ASME's fiftieth anniversary celebrations. In fact, the story also reflected a set of relations that in 1930 were real and important. *American Machinist* was but one of many technical journals published by McGraw-Hill. Mechanical engineers were the prime customers for it, for *Factory*, and for *Power*; this market by itself would probably have warranted sponsorship of the entertainment and luncheon McGraw-Hill presented to the delegates on 5 April. Beyond business interests, however, there was still another important connection between the Society and the world of technical journalism. Senior editors filled leading posts in the Society and, together with staff members of the large engineering and technical organizations based in the city, were part of an 'invisible college' dedicated to the promotion of engineering in America and of New York as its center.

A relation very much like that between ASME and McGraw-Hill also existed between the Society and New York area engineering schools, particularly Stevens Institute of Technology, across the Hudson River in Hoboken, New Jersey. The organizational meeting in April 1880 had taken place in the school's auditorium, and like that editorial office gathering had taken on mythic qualities. But in a similar fashion, too, the connection between institutions had significant practical side to it as well. The Society's first president, Robert H. Thurston, was a Stevens faculty member; Alexander C. Humphreys, president of Stevens from 1902 to 1927, was also president of the Society, as was his successor at Stevens, Harvey N. Davis; David S. Jacobus from Stevens was also president, and others filled highly important committee positions in ASME for long periods of time. Thus there existed a vigorous and mutually rewarding set of relations between the two institutions. The Society depended heavily on nearby members for leadership in certain crucial activities and Stevens faculty members took on many of those tasks. Their efforts, in return, enhanced their professional standing and the reputation of the Institute to the benefit of its enrollment and endowment. It was for these kinds of practical reasons, as well as sentimental attachment, that the Stevens Board of Trustees pledged $10,000 toward the production of another fiftieth anniversary pageant which would link mechanical engineering, ASME's history, and the progressive development of civilization.[6]

That was a substantial amount of money, even in the bull market optimism of the spring of 1929, and the school's board of trustees must have expected some sort of return on their investment. Furthermore, it deter-

6 Minutes of the Board of Trustees, Stevens Institute of Technology, 3 April 1929; Stevens Institute, Hoboken, NJ

mined to have the best pageant money could buy. To write and produce it, the board commissioned George Pierce Baker, chairman of the drama department at Yale, who had achieved great popular fame in America for his pageant at Plymouth, Massachusetts, to commemorate the 300th anniversary of the Pilgrims' landing.[7] Baker called his production for Stevens *Control: A Pageant of Engineering Progress.* The first part of the title was not meant to be ominous, but to suggest Thomas Telford's eighteenth-century definition of engineering as 'control over the forces of nature for the benefit of mankind.' Far, indeed, from any dark implications, the pageant aimed at a triumphal portrayal of engineering progress, an approach that perfectly complemented the sober theme of the Washington lectures, 'The Influence of Engineering Upon Civilization.' The theatrical form was just right, too. Traditionally suited for historic occasions, pageants were familiar and emotionally rewarding. But *Control* went far beyond the normal reach of such affairs. To a remarkable degree, the pageant at Stevens Institute dramatized the fondest images engineers had of their profession and it captured their hearts as practically nothing else during the remainder of the week.

Control was the last of Baker's pageants, but it mixed together ideas and elements characteristic of his work from the beginning. His pageants were dominated by symbolism and allegory, yet they also depended heavily on historical texts for the spoken word. And despite the fact that he imagined the pageant as an ideal theatrical form for small communities – a sort of democratic vehicle with which to celebrate local history and strengthen local pride, costly special effects were also the hallmark of his work.

Set on the beach behind the famed rock, the Plymouth pageant had been Baker's most spectacular. Expensive, grand in conception, and technologically elaborate, it aimed to link the spirit motivating the Pilgrim emigration to America with broadly patriotic themes. The concluding scene epitomized Baker's style: the *Mayflower* now rode securely at anchor in Plymouth harbor and a hidden choir sang Robert Frost's 'The Return of the Pilgrims,' while a procession of young women bearing the flags of the forty-eight states came into the amphitheater to join the cast on stage. As the choir's last notes died away, specially devised spotlights came up full on the *Mayflower* and searchlights swept the skies. Then the lighting dropped away until only the ship stood in a single spotlight and from a loudspeaker near the famed stepping stone, the 'Spirit of the Rock' spoke for the final time: 'With malice toward none and charity for all it is for us to resolve that this nation under

7 Wisner Kinne *George Pierce Baker and the American Theatre* (Cambridge: Harvard University Press 1954)

God shall have a new Birth of Freedom,' and then the last spotlight slowly faded out on the Mayflower.[8]

Control seemed to offer no such scenic possibilities. Instead, the limitations placed on Baker read like a set of technical specifications. The pageant was to be held indoors and was restricted to a small stage in the auditorium of Stevens Institute, it could last only an hour, and all performances had to be on the same day. Baker turned these limitations to advantage: 'Throughout, this Pageant will differ greatly from any Pageant heretofore given, in that it will depend as little as possible on acting and as much as possible on recent inventions. Throughout, accompanying music, to be carefully selected, will be from the Electrola or Radiola. No band or orchestra.'[9] He subsequently added motion pictures and the projection of slides to round out the idea that a pageant about engineering ought to feature the latest technology. It was 'quite an amazing conceit,' assistant stage manager Edward Cole recalled, and one that worked remarkably well in practice.[10] Others who were connected with the pageant also remembered it long afterwards for the novel special effects Baker achieved with sound and film.

There was much in the pageant, however, that depended on concepts he had worked out previously. Allegorical figures, a favorite Baker device, were used to link historical episodes, just as in the Plymouth and earlier pageants. Thus, *Control* opens with films of landscapes and seascapes, then rivers and waterfalls, storms, geysers, and volcanoes. As primitive peoples, who are gathered on each side of the stage, react with fear and religious awe to these violent natural phenomena, the shrouded symbolic figure Mystery 'appears on stage gropingly' from among a group of people worshipping about a totem pole. 'I, Mystery, the Mole of Nature,' the figure says, 'grope blindly and helpless; seen, wondered at, unknown, unsolved. Who shall master me, revealing my secrets?'[11]

As Mystery crouches at center stage, fully covered with his robes, another allegorical figure, Curiosity, emerges from one of the groups of primitive people, followed by Intelligence and then Imagination. Each in turn speaks a few lines to explain his nature and the three then join hands to circle around Mystery and bring 'from beneath the folds of his robe the figure of Control as a child of ten or twelve,' who comes in front of Mystery to say: 'Thus

8 Ibid 225–6
9 George P. Baker to James Creese, 5 Aug. 1929; Stevens Institute Archives
10 Personal conversation
11 All the descriptions and quotations from the pageant are taken from the published version, George Pierce Baker *Control A Pageant of Engineering Progress* (New York: The American Society of Mechanical Engineers 1930).

Mystery, searched by Curiosity, Growing Intelligence, flashing Imagination, is stripped of all its mystery, revealing me, Control – a child as yet.' 'Child Control' represented the early stages of engineering, and in another of the pageant's conceits, the role was played by a descendant of John Stevens, for whom the institute was named.

The first historical scene then depicted James Watt, in his workshop, soon joined by Matthew Boulton. Watt is in despair for lack of funds with which to develop his steam engine. 'How I long for you as an understanding partner,' he says to Boulton, 'fully able to make all I need possible.' As the scene unfolds, the audience learns that Boulton's wish to help his friend is frustrated by John Roebuck's prior interest in the steam engine. But when Roebuck fails, the means are at hand for Boulton to take over, which he quickly does, promising Watt all the facilities of a first-class Birmingham workshop, in a speech that ends with a version of that famous remark 'we will sell what all the world desires to have – POWER.'

The central message of this first historical episode was that invention was helpless without capital, and a sort of allegorical entr'acte recapitulated it. Youth Control, now grown to a young man 'fine and strong in figure, particularly intelligent in looks,' is joined by Finance, who says 'and so Control wins to his aid Finance. Each of us apart – inept and ineffective; together – trained, unconquerable.' Youth Control eagerly accepts Finance to the company of Intelligence and Imagination:

All Nature shall be mine to conquer – all! Growing Intelligence, richer Imagination urging me on, Finance supporting me. So march we four to marvels unpredictable *(Mystery now flees in terror)*
Ours first to conquer
Incredulity and prejudice
In men more willing to trust
What is and has been
Than to understand our dreams of what may be ...

The last phase was calculated to strike a responsive chord, reminding the audience of the inscription on the fiftieth anniversary plaque, 'what is not yet may be.'

The distinction between those mired in the past and those looking to the future also served to introduce the next historical scene, in which George Stephenson appears before a House of Commons committee to testify in favor of his proposed railway between Liverpool and Manchester. He is told the line would threaten the interests of coachmen, innkeepers, and harness makers; that the speed and noise of his machines would disturb the peace of the countryside; and, in the kind of jokes on history engineers liked to hear, that the construction of rolling stock would exhaust the country's iron resources while the speed of twelve miles an hour was beyond human capabilities. Stephenson is taunted for his lack of formal education, but the dialogue makes clear that besides indomitable conviction, experience has provided him with substantial technical knowledge. And the morality of his undertaking is defended with the prediction 'the time is coming when it will be cheaper for a working man to travel upon a railway than to walk on foot.'

As the curtain falls, the allegorical figures return. Youth Control points out that Stephenson succeeded in his ambitions, and Finance reports that the railway came in under estimate, while Intelligence and Imagination look for new fields to conquer. 'Yes,' Finance says, 'with my aid, the lightning's flash shall yield to you, revealing all the mysteries of the air.'

The pageant then moves to the Ashmolean Museum and Faraday's successful demonstration of electromagnetic induction. The issue of this episode is simple; against all urging, Faraday refuses to pursue both the obvious financial wealth that his discovery would bring and the honors that would be heaped upon him. Instead, he claims, 'My business is to discover unhindered and uncomplicated, and then to defend,' an argument he repeats, in the only dark reference of the pageant, when it is suggested his discovery puts 'new arms into the hands of the incendiary.' Against that threat, the allegorical figures proclaim the benefits of telegraph and telephone, radio, and, in the presence of a new symbolic figure, Conversion, of electric power.

Conversion represented both the use of steam power to generate electricity and the historical emergence of a new level of technical sophistication, and to make the last point plain, Baker utilized flag-bearers again, this time with the banners of America's technical schools, brought in procession down each side aisle to the stage. However, as Imagination explains, schools did not supply the need for the exchange of information among practitioners and for their fellowship in 'common service for science.' Thus, the audience was provided with a connection to the next historical episode, depicting the first regular meeting of the American Society of Mechanical Engineers which had been held in the same Stevens Institute auditorium fifty years earlier. Drawn from Frederick Remsen Hutton's 1915 history of the Society, the scene is meant to portray the good companionship and sound judgment of the founders in framing membership and election policies.[12]

12 Frederick Remsen Hutton *A History of the American Society of Mechanical Engineers from 1880 to 1915* (New York: ASME 1915) 19

As the curtains close on the firm establishment of the Society, Youth Control is replaced, 'if possible unnoticed,' by Mature Control. 'Steady of figure and clear of eye,' Mature Control symbolizes both the technical and the professional coming of age of engineers and his speech indicates still new arenas calling for such skills: 'All reveals that in every age engineering in some form has been one of the fundamental means by which civilization has advanced. Engineers are becoming a controlling force in culture, politics, commerce, industry, finance, education, and national defense.'

The final historical sequence suggested the still unfolding potential of systematic engineering thought, as well as its awesome potential. The scene centers on Edison's demonstration of electric lighting. A crowd gathers before a building Edison has announced will be lighted by electricity at a given hour. Their initial scepticism turns to wonder as the lights come on, and the stage directions indicate they should assume poses like those of the primitive people at the pageant's beginning, when they were confronted by natural forces beyond their understanding. This gulf between their knowledge and Edison's is underscored by a sequence in which a bumptious reporter who questions Edison would trivialize the accomplishment because of his own ignorance. The mood then turns serious and one of the bystanders asks Edison what he considers his real work. Reiterating the idea behind the pageant's title, Edison answers, 'Why – why – bringing out the secrets of Nature and applying them for the happiness of man.'

As Edison hurries back to his laboratory, the crowd is left to marvel at the implications of this new technology, the curtains close, and the symbolic figures return to the front of the stage. They chant a chorus of engineering accomplishments in a litany that increasingly emphasizes the words 'power' and 'beauty,' while films are shown of automobiles, trains, steamships, and skyscrapers. Then, in a splendidly fantastic conclusion, and 'in a great glow of light and color,' Beauty, the last of the allegorical figures, emerges. She, it turns out, is the child of Control and Imagination, and it is she who has the last prophetic line: 'I beckon ever into greater heights and flights.' A grand march brings all the players on stage, and in a fine Baker climax, President Hoover appears on the movie screen with a personal message for the occasion, fading out to 'America the Beautiful' which the entire cast joins in singing, while 'colored lights flash and whirl over all, dying as the curtains slowly close and the pageant ends.'

Control was a great success. Baker had brought into the production practically the entire Yale drama department, the leading department in the country. Stanley McCandless was already famous as a lighting expert, Fred Bevan would go on to Broadway success as a costume designer, and the music and scene design jobs were handled by people of equal stature in the field. Even Harold Burris-Meyers, a new member of the Stevens faculty who worked on the sound effects, discovered a subsequent career in the use of acoustical techniques to modify behavior that led him to a vice presidency of Musak. The institute gave him a $250 bonus for his extra work on the production. As 'Master of the Pageant,' Baker's fee was $5,000 while Edward Cole, his assistant at Yale, joined the troupe as assistant stage manager at a figure well beyond his expectations. McCandless's lighting plans also required extensive modifications to the Stevens theater. And the crew lived well, spending their evenings at a German restaurant in Hoboken that continued to serve good beer throughout Prohibition.[13]

Much of the impact of *Control* came from the crew's skill in devising new and imaginative combinations of lighting, sound, and film. The musical passages were carefully selected and perfectly synchronized, while the films had a particularly stirring effect on the audience. One strip of footage, showing the new steam-turbine ocean liner *Europa*, was especially exciting since the ship had just broken the transatlantic speed record and the pageant's films of it were the first shown in America. But there was still more to it than that. Baker always claimed that pageants had to have their 'own particular core of meaning,' and for *Control* he found it in the vision engineers had of their history.

Baker had taken two of his sketches directly from Frederick Hutton's history of ASME and depended on it also for general background and tone. He worked closely with James Creese, vice president of Stevens, on the rough draft for the pageant as well as on subsequent revisions of it, and it is easy to imagine that Creese supplied Baker with other elements of engineering history. The dramatist, however, also had his own perception of America's mission to the world, amply expressed in previous pageants, and it blended nicely with the view engineers had of their historic role in the development of civilization. The pageant therefore gave its audience of engineers the rare chance to see their own images of the profession fully articulated. And Baker's skillful dramatization of the Society's most familiar mythic elements, set against the background of mankind's progress from superstition to science, made them all seem real.

Most of those who sat in the audience were successful engineers and prominent members of ASME and that may have made it easier for them to believe that what they saw had a vital nucleus of truth to it. Their own careers proved that foresight and determination were rewarded, that the

13   These details come from conversations with Harold Burris-Meyers and Edward Cole.

combination of capital and technical skills was powerful, and that the Society was led by men of vision and good fellowship. They could see themselves in Control's sure growth from lusty youth to vigorous manhood, just as they could see their Society's evolution to a plane of statesmanlike professionalism. Baker's pageant made all those ideas suddenly alive; ASME's history was their own history and it was splendid to have it portrayed so compellingly.

The Washington ceremony, which began on Monday, 7 April 1930, presented an equally unusual opportunity to see demonstrated the dignity and prestige of the profession. The planning committee's basic idea was to have the major geographical divisions of the world represented by an outstanding engineer, who would speak briefly about the influence of engineering on the development of his country. Each delegate would be introduced by his country's diplomatic representative in Washington and, to add extra color to the occasion, would present his country's greetings in the form of an engrossed scroll. And each foreign delegate, for his own outstanding contributions to engineering, would be awarded one of the Society's specially struck fiftieth anniversary gold medals. Beyond that remarkable array of expertise and diplomatic pomp, the Committee on Meetings and Program had secured Robert A. Milliken, 1923 Nobel laureate in physics, as the after-dinner speaker at the anniversary banquet. And best of all, a new gold medal for outstanding public service by a member of the engineering profession was to be presented that same night. The funds for it had been provided by Conrad Lauer, a Vice President of ASME, but in a generous gesture befitting the occasion, it was to be awarded in the name of all four of the country's major engineering societies. Called the Hoover Medal, it would be presented the first time to President Hoover himself.

Each element of the ceremony met the committee's keenest hopes. The format imposed on the foreign delegates might easily have produced a stultifying series of generalizations. Instead, the talks were full of variety, interest, and even candor. The Austrian delegate pointed out the relation between his country's great loss of territory as a result of the war and the kind of engineering projects that consequence dictated. Belgium's delegate laid most emphasis on his country's programs of industrial health and safety. Conrad Matchoss, technological historian, the distinguished director of the *Verein deutscher Ingenieure*, and a person already known to many in ASME, addressed himself directly to the human problems of mechanization. He claimed that 'in spite of the injurious effects of the so-called machine age, there could be no turning back therefrom.'[14] In high-voltage transmission lines, he saw the 'con-

14 As quoted in *Mechanical Engineering* 52 (1930) 518

structive spirit of engineering progress' flowing without hindrance across national boundaries, but that kind of progress had an inevitable quality about it and Matchoss predicted adaptation to that new world as mankind's only hope.

Loughnan St L. Pendred, editor of *Engineer* and president of Britain's Institution of Mechanical Engineers, was also known to many in the audience and he also spoke to the Machine Civilization issue. In a way that only a person of his standing and accent could do, Pendred expressed engineering's grandest ideas for solving the human problems of modern society. The influence of mechanical engineering had not always been a positive one, he admitted, particularly in its effects on home life. But he called upon his fellow engineers to dedicate themselves to a 'new and greater service' than the satisfaction simply of material needs. They, and they alone, had 'the power to make war or preserve peace;' through their technical skills, applied to land, sea, and air, engineers could 'knit the whole world into a single unit,' provide the abundance to eliminate envy, and the leisure to fructify the arts.[15]

Even given their eminence, Pendred's audience scarcely had the power to bring about the political and cultural revolutions he urged on them. Yet, throughout the celebration, engineers were encouraged to think of themselves as the agents of major change in human affairs – true revolutionaries instead of those who only talked about reform. Robert Milliken struck the same note in his address at the fiftieth anniversary banquet. Scientists and engineers were uniquely responsible for the upward curve of mankind's history, he argued, and the special training and method that had so successfully harnessed the forces of nature were now required for social and political leadership. Milliken's message may have sounded much as others before it, but it had particular value and importance to the engineers who heard it. Unlike many scientists known for their theoretical contributions, Milliken actively promoted industrial research and the practical applications of science. Besides, he was a staunch supporter of the existing economic order.[16] In any event his ideas would have been warmly received, but on this special occasion they came wrapped in the prestige of America's premier physicist.

The culmination of the banquet was, of course, the presentation of the Hoover Medal. Dexter S. Kimball, a past president of ASME, described how the award had come about. At the beginning of the twentieth century the engineer had been regarded by the general public simply as a technician, a person inappropriate to deal with 'matters pertaining to civic welfare and

15 Ibid 519
16 Daniel J. Kevles 'Robert A. Milliken' *Scientific American* 240 (1979) 124

public service.' But Kimball claimed that a great change had taken place over the past thirty years and he noted that, just as Milliken had pointed out before him, 'a chance for service equal to that of any of the great learned professions' was now open to engineers.[17] To mark that important change and because there had been no awards particularly to honor distinguished public service, the Hoover Medal had been created. In the seriatum style of introduction favored on such occasions, Kimball then presented J.V.W. Reynders, a past president of the American Institute of Mining and Metallurgical Engineers, Hoover's own professional society, that he might introduce the President.

Reynders cast his introduction in an interesting form. He set current problems and issues against the time fifty years later when ASME would be celebrating its one hundredth anniversary. Would America's population, he wondered, still have its Anglo-Saxon characteristics? Would the country's social structure have equitably satisfied its citizens' reasonable ambitions? Would it to any serious degree have felt the effects 'of the diminishing reserves of our unexampled natural resources, of oil, of copper, of lead?'[18] Fifty years from now, Reynders said, the nation's leadership would face other problems and it was difficult to imagine what they might be. Nonetheless, he claimed, with President Hoover's example before them, engineers could be counted on to play a role of increasing importance in the resolution of those difficulties.

Hoover's remarks were brief and appropriately modest, but directly upon the theme of the past four days. America's technical capability – 'our great national tools' as he put it – had brought manifold benefits to the country, but also made it necessary for government to insure that the control of technology was not misused to diminish equality of opportunity or to restrict freedom. Modern development had created a host of what Hoover called 'public relationships' to industry that required technical knowledge to solve them. Hoover saw these new problems of government as essentially technical in nature, however, not political. It was a matter, just as in engineering practice, of determining the facts, setting them in perspective, and applying practical experience. Unlike engineering problems, these public relationships involved 'emotions, that confused the data and excited controversy,' Hoover said, but that was just why the cool rationality of the engineering method was so much needed. Not only did engineers have an important

17 Stenographic transcript, Fiftieth Anniversary Banquet, 8 April 1930; ASME, New York, NY
18 As quoted in *Mechanical Engineering* 52 (1930) 527

contribution to make in the public's welfare, Hoover concluded, but they also 'have an obligation to give that contribution.'[19]

Those who were there described the evening as 'an almost perfect affair,' or, less equivocally, as that 'perfect Anniversary Dinner.' And so it should have been. Held in the grand ballroom of the Mayflower Hotel, with 800 delegates, ambassadors, ASME members, their wives and guests, all in full evening dress, the banquet had begun with trumpet flourishes and the Marine Band playing 'Hail to the Chief' as the President and Mrs Hoover took their places at the head table. William F. Durand, a past president of the Society whose biography of Robert H. Thurston, its first president, had appeared the year before, proved an excellent toast-master, urbanely striking just the right balance between good humor and the kind of rhetorical fulsomeness his task called for. Besides the Society's distinguished guests, the head table also included old and familiar figures such as Calvin W. Rice, ASME Secretary since 1904 and closely associated with major changes inside the organization, but also widely known abroad for his efforts on behalf of engineering internationalism. And through the efforts of Rice's assistant, Clarence E. Davies, the dinner was also characterized by the high degree of organization that had been evident throughout the fiftieth anniversary celebration. In a profession where foresight and systematic planning are cardinal virtues, those responsible for the details worked hard to insure that the occasion measured up.

Everything about the enterprise, in fact, was calculated to make it noteworthy. Each person in attendance received an individually numbered, leatherbound, pocket-sized guide and program to the fiftieth anniversary events plus a special anniversary edition of *Mechanical Engineering*. There were special buses in New York, special trains to Washington, and special tours. Tickets were arranged, luggage looked after, and two extra rooms were set aside at the Mayflower as the 'Ladies Headquarters.' Even the White House, the day after the banquet, was opened for a special reception. The physical arrangements, in fact, mirrored one of the central messages of the anniversary, that in the judgment of history mechanical engineers rightfully occupied a place of crucial importance in modern society.

There is nothing in the Society's official records to suggest that the planning committee approached the anniversary with an explicit set of public relations objectives. They had a general approach in mind and some sense of the level at which they wanted things carried out. But other than the theme 'The Influence of Engineering upon Civilization' there were no directives that might have imposed unity on the program, and there were no sub-committees to insure co-operative action among its separate parts. Yet the elements combined with a remarkable coherence. Baker's pageant, for instance, might well have been a thing apart from the rest. After all, it was commissioned by Stevens Institute simply to commemorate an occasion, and aside from the co-ordination of timetable, ASME's committee apparently had nothing to do with it. Besides, Baker's temperament, the nature of the art form, and the event itself all dictated an up-beat approach, without any need for direction from someone else. But as it happened, *Control* struck notes that continued to be sounded throughout the remainder of the celebration. In fact, all the parts of the program that began on 5 April in New York, deliberately historical in nature and traditional in form, perfectly complemented those subsequent Washington events later in the week which were meant to be factual and contemporary in mood. Indeed, the two major parts of the anniversary blended together so well one might imagine the committee had actually planned it.

It was, however, the 'Machine Civilization' debate, sharpened by economic depression, that gave ASME's fiftieth anniversary such thematic consistency, and what emerged from the week-long celebration was an ideology with which engineers meant to defend themselves against their critics. As the editor of *Mechanical Engineering* pointed out, 'Engineers are keenly conscious of the strategic position they occupy in the modern world. They recognize their obligations, and they resent much of the criticism directed at them and at the so-called machine world they have helped to create.'[20] But it was one thing to orchestrate the Society's mythology for the purpose of the fiftieth anniversary celebration and quite another to convince a larger, non-engineering, audience that the historical lessons drawn from it were true. After all, according to their own self-image, engineers were 'more used to deeds than to words' and thus less able to defend themselves against the attacks of their 'literary-minded critics.'[21] In fact, however, as the planning for the anniversary was under way, several of the leading figures in ASME were organizing just such a response. Convinced in their own minds of the value of their profession's approach to the country's current problems and determined to answer the critics of mechanization, they looked upon the anniversary celebration as an important opportunity to present 'the engineer's point of view.' And once more, they chose history as the vehicle for

20 Ibid, section two, n.p.
21 Ibid

their defense. Or, to put it more accurately, they chose a historian, Charles A. Beard.

Beard's own sense of American history predisposed him to enter the Machine Civilization debate. In his view, the extended conflict between an older agrarian society and an emerging aggressive industrial America was but the leading edge of a similar historical movement throughout the rest of the world. For Beard, science and the machine were at the heart of modern civilization: that idea shaped his *Rise of American Civilization*, written jointly with his wife in 1926, and it led him the following year to *Whither Mankind*, an edition of essays that brought him firmly down on the side of the new age.[22] ASME appealed to Beard to edit a sequel in which engineers, whose voice had been entirely omitted in *Whither Mankind*, could defend their work. The result, published in 1930, was *Toward Civilization*, an attempt more explicitly and rationally to argue the case *Control* had so well dramatized: that modern society rested upon the engineer's accomplishments and that he was now ready to apply the same skills in the solution of wider problems.[23]

Many of those who had been prominent in the anniversary celebration participated in writing the book. Milliken contributed his banquet speech retitled 'Science Lights the Torch'; Dexter Kimball wrote on 'Modern Industry and Management'; Harvey N. Davis, president of Stevens Institute, wrote the concluding chapter, 'Spirit and Culture under the Machine.'

But the leading spirit in the enterprise was Ralph Flanders, an ASME Vice President, a New England machine-tool industrialist, and a man with political ambitions. He wrote the lead essay 'The New Age and the New Man,' and it followed along lines strikingly parallel to ideas that Baker had dramatized in his pageant. Beauty, for instance, had figured prominently in *Control*, suggesting a final stage of professional development and defining a new level of engineering concern and achievement. Her surprising introduction into the pageant at its very end was meant to counter the indictment that mass production culture was esthetically sterile, even destructive to artistic capabilities. That charge was a central element in the attack on Machine Civilization, and Flanders focused his attention directly upon it. First, he claimed that the machine era had the potential for an inherent beauty of its own, even if obscure to the unsympathetic. 'Most attempts of artists to give expression to our age,' Flanders said, 'seem crude, ridiculous, uncompre-

22 Beard ed *Whither Mankind*; David W. Marcell *Progress and Pragmatism: James, Davey, Beard, and the American Idea of Progress* (Westport, Conn.: Greenwood Press, 1974) 275
23 Charles A. Beard ed *Toward Civilization* (New York: Longmans, Green 1930)

hending and insincere.'[24] Instead, the search for beauty should concentrate on 'fitness of line, mass, and coloring based on structure and use.' Flanders then described, in a catalog remarkably like the set of pictures Baker projected in the final scene of *Control*, what he meant by this kind of direct and simple beauty:

The fitness of the racing yacht; of the bow and superstructure of the *Bremen*; of the New York Telephone Building; of a Cincinnati Milling Machine; of the stream line body of the Renault 'Monosix'; of the sturdy self-respecting splendor of the *Kungsholm* (*à bas* the *Île de France* with its neurotic feminism); of an airplane; of the sleek electric train, gliding serpent-like up the grade to the St. Gothard; of the Philadelphia-Camden bridge; of the apparatus of the modern bath-room. All this fitness is of the integral, organic sort in which the engineering sense is the determining factor.[25]

Beyond his own canons of beauty, Flanders imagined a connection between the productivity of the machine era and a whole range of artistic possibilities. To begin with, mass-produced items themselves could be better designed; the 'New Age' would demand 'they have this *fitness* of line, form and color.' Furthermore, the higher standard of living from mass production, according to Flanders, would lead to the revival of craftsmanship, but this time at a level of 'dignified support and assured position.' That same affluence would also provide time for the study of beauty, creating from an analysis of past and present 'a beauty that is truly modern.'[26]

All these hopes were allied to a sense in the 1920s that the world might in fact be on the verge of a new stage in human history. It seemed possible to relegate the industrial horrors of the nineteenth century to an outmoded technology of steam power. Electricity promised cleaner cities, a new approach to factory organization, and an enhanced degree of control over the productive processes. But most persuasive to the vision of Flanders' 'New Age,' Lewis Mumford's 'Neotechnic Phase,' or what Walter Polakov called 'The Power Age' was the unchallenged ability of the modern machine system to produce a level of abundance so much greater than any period of the past that it suggested a historical disjuncture.

And that was ASME's fiftieth anniversary message; abundance proved the value of engineering contributions to the world. Machine Civilization put nature's resources at the disposition of more people, allowing them to travel

24 Ibid 31–2
25 Ibid 32
26 Ibid 32–3

faster, build higher, and live with greater material comfort than ever before. Furthermore, it was in just such dramatic historic contrasts that the evidence told most clearly. Now ordinary folk could daily afford what had once been the luxuries of kings alone. The age-old curse of burdensome toil had finally been lifted, and even the average working man now had leisure to re-create himself. It was as if all of mankind's history could be lumped in two main divisions. as Baker's pageant did, one before the steam engine and the other afterwards. In that vision of the past, technical achievement determined human progress. It is also worth noting that this view of technical achievement emphasized values of speed, size, and energy. Its primary constituent elements were electric power generation and mass production technology – two main concerns of mechanical engineering.

This built-in set of assumptions about technology, however, was a practically invisible element of the ideology that emerged from ASME's fiftieth anniversary. Not that alternative ideas did not exist; there were those who claimed that a smaller scale of technology was possible – that electric motors, for instance, might be used to revivify the home manufacture of things and thus give a more human scale to industrial production – but they were not involved in the celebration.[27] Instead, the theme of abundance implicitly argued the value of a large, highly organized, capital intensive style of technology, a system in which the centrality of mechanical engineers was evident.

There were other images that ASME's leaders explicitly used the anniversary to establish. One had to do with the qualities of the mechanical engineer. Besides the personal characteristics *Control* so well mythologized, engineers were persistently identified during the Washington sessions with objectivity and a kind of occupational neutrality. Charles Piez, the Society's president, made the engineer the crucial agent in the transformation of inventions into products serviceable to mankind, while Milliken described the position as intermediary in the conversion of scientific ideas to practical utility. In either case, the transforming process involved rationality, systematic procedure, and balanced judgment. These anniversary job descriptions bore little resemblance to what most members of ASME actually did, but it was not the function of the celebration to resolve such differences.

The anniversary also proclaimed that the American Society of Mechanical Engineers had reached a new level of professional consciousness. Mature *Control* symbolically represented the Society's arrival at that stage of deve-

27 See, for example, Ralph Borsodi *This Ugly Civilization* (New York: Longmans, Green 1929).

lopment and the idea found repeated expression. Furthermore, it was generally suggested that the metamorphosis was practically coeval with the fiftieth
anniversary. The 'then' and 'now' style of historical contrast used to describe
material progress was employed in this case, too. Compared with its earlier
history, when the organization limited itself to technical matters only, the
Society was now ready to assume its responsibilities in public affairs. And
that idea carried with it the conclusion that engineering training and methods
were the best kind of weapons to solve the social and economic problems of
modern industrial society. Indeed, Hoover had said these were not solvable
without the engineer's approach to truth.

In the spring of 1930, before a worsening depression rendered the dream of a
new age of abundance somewhat abstract, before Franklin D. Roosevelt and
his disturbing programs made public affairs so distasteful, it was possible for
some of ASME's principal figures sincerely to believe that the moment had
come for engineers to assume a leading role in America. Certainly Ralph
Flanders thought so. It seemed time 'to manhandle our civilization and
make some intelligent effort to control it.'[28] But first it was necessary to
convince the country that engineers had all those special and relevant qualities they claimed to have. That did not promise to be an easy job, since so
many of those who helped form public opinion, with backgrounds in the
humanities and social sciences, were unsympathetic to the idea. The only
answer, *Mechanical Engineering*'s editor George A. Stetson thought, was 'to
keep hammering on the engineers point of view.'[29]
   In a way, of course, that was the point of the whole enterprise – Baker's
pageant, the Washington meetings, and *Toward Civilization*. Together they
constituted one of the most elaborate, broadly gauged, highly organized, and
skillfully worked out public relations campaign any American engineering
society had even undertaken. And rarely had any professional group so concerned in their daily working lives with the present ever devoted so much
attention to the past. Convinced that in the sober judgment of history
mechanical engineers would not be found wanting, those who directed the
Society's affairs were nonetheless unwilling to leave the decision to time.
Their economic judgments, their sense of social order, and their style of
technology were all under attack. What they shaped, therefore, was not an
analysis of their past. Instead of self-knowledge they fashioned an ideology
they could wield in the struggle with their enemies.

28 George A. Stetson to Ralph E. Flanders, 21 Nov. 1929; Flanders Papers, Syracuse
    University, Syracuse, NY
29 George A. Stetson to Ralph E. Flanders, 20 May 1930; Flanders Papers

Like a flag carried into battle, the keynote of 'progressive optimism' served as a rallying cry for the fiftieth anniversary. But the set of images it meant to evoke – of a Society secure in its sense of accomplishment and unified in its ambitions for the future – disguised a welter of problems. Contrary to the language of the celebration, for instance, ASME was not at all resolved in the matter of public affairs. That vexing issue had agitated the organization for at least two decades and it was not settled yet. Indeed, the general question of social responsibility for the profession sheltered a number of related but separate questions, all of them exacerbated by the depression and the assaults on mechanization. Was the organization principally to serve technical interests? That is, should it concern itself primarily with the creation and diffusion of technical knowledge, particularly the type of value to the industries that employed mechanical engineers? Or, to what degree was the Society primarily responsible for the broad social and economic interests of the membership at large, in the manner of a guild? Was it appropriate for a professional society to take a politically active role, to lobby for legislation reflecting the engineering point of view? And how could a membership so widely dispersed over the country feel a connection to ASME's New York headquarters, especially when it seemed that a core of metropolitan members exercised disproportionate power in Society affairs?

The fiftieth anniversary celebrations did not speak directly to any of these issues and no one expected otherwise. It would have been completely uncharacteristic of the profession to make such questions the subject of public discussion, particularly on an occasion that so clearly called for traditional styles of observance. But as it turned out, all that attention to the past did not serve the Society very well either. In a fashion not uncommon among professions, the engineers confused history and mythology. And because their perception of the past was damaged, they were limited in their ability to deal effectively with the real problems they faced. Thus ASME's fiftieth anniversary provides an example of the real meaning of Santayana's familiar observation that those ignorant of their history are compelled to relive it. Those unresolved issues of 1930, masked by the ideology of 'progressive optimism,' would remain for the Society to confront in other times.

2

Professional Men of Outward Success

Thirty of the most prominent men in American mechanical industry attended that first meeting of ASME founders in the New York editorial offices of *American Machinist* on 16 February 1880. They chose as chairman the brilliant consultant to the American Bessemer Steel Association, Alexander Lyman Holley, and, characteristically, he provided a focus for the gathering, outlining both the intellectual boundaries of the mechanical engineering profession and the advantages to be derived from association. All the steps necessary to establish a new engineering society were taken at that meeting. It generated a membership list, committees to nominate officers and to draft by-laws, and scheduled a formal organizational meeting for 7 April in the Stevens Institute auditorium to ratify these measures. The number of those interested in the project grew over the next two months. About 80 men attended the second meeting, which among other things, established for years to come the exclusively masculine character of the Society.

The Stevens assembly also adopted rules defining the new organization's governing structure. Besides a president and treasurer elected annually, the officers consisted of six vice presidents, each serving two years, and nine managers, each elected for three years. Their terms were arranged so that of the fifteen, one-third rotated out of office each year, providing both a sense of change and of continuity. Although it was never explicitly stated, those framing ASME's governing procedures imagined that presidents would be chosen from the ranks of vice presidents, who in turn, would usually be selected from among the managers. This group of officers comprised the Council, which at first carried out many of the functions involved in program planning and in publishing the Society's *Transactions*. In fact, until 1904, the Council also served as the organization's membership committee.

From the outset, the Society had an intimate character. The number of members was small and the leading figures were remarkably alike in background. Alexander Holley had purposely balanced the first Council to represent the major branches of mechanical engineering, but the dominance of machine builders in the early years was clear. They included those preeminent in the design and construction of pumping engines, such as Erasmus Darwin Leavitt and Henry R. Worthington; Coleman Sellers of William Sellers and Company and Francis A. Pratt of Pratt and Whitney, two of the country's outstanding machine-tool builders; and such well known power plant constructors as John E. Sweet and Charles T. Porter, both of whom had developed steam engines especially suited for generating electric power. Thus from the first, the organization was shaped by the style and concerns of men who possessed great mechanical ingenuity and considerable business talents, men who had come from the machine shop floor even though, more often than not, their families owned the firm.

The closely knit quality of ASME leadership influenced the way they did such things as selecting the secretary, an administrative position created at the Stevens meeting. Two of the first three men to fill the position were related to prominent members. Thomas Whiteside Rae, for example, Secretary from 1880-82, was Henry R. Worthington's son-in-law. The founder of the Worthington Pump Company and a Vice President of the Society, Worthington also gave the enterprise office space in his New York headquarters. Frederick Remson Hutton, an assistant professor at Columbia University and a person of some independent means, whom the Council appointed Secretary in 1883, at first carried on the job with the help of his brother in a downtown office he rented himself. But the Society soon found him space at its own expense and, because of a growing administrative work-load that went beyond clerical detail, he became the first secretary to exercise executive functions.

Hutton thus played a double role in the Society, as did subsequent secretaries. He had much of the responsibility for insuring that ASME's annual and spring meetings (which in time became the winter and summer annual meetings) went well. That not only meant trying to get good papers to be delivered, but arranging for prepared discussion by engineers competent in that particular subject, and then converting both into the Society's *Transactions*. Professional meetings and publication had early been identified as the most important of ASME activities, and the secretary's success inevitably came to depend on his own leadership qualities. Yet the Council exercised the actual power in the organization and it was filled by men long accustomed to authority.

The Society held its first annual meeting in New York in early November 1880 and then, with an occasional exception, settled into a pattern of annual meetings in New York, where its headquarters were, and spring meetings in others parts of the country. The pattern of meetings also became routine fairly quickly, as did the format of annual volumes of ASME *Transactions*. The relatively smooth transition from an idea for an organization to one that soon fell into easy patterns of operation suggests that the community which called it into existence had a clear sense of its own needs and aspirations.

THE PHILOSOPHICAL FOUNDATIONS OF ASME

When the American Society of Mechanical Engineers' beginnings were recalled at the fiftieth anniversary celebration, the founders were depicted as solid and skillful men, who recognized the need for a better system of exchanging technical information and looked for it in a social setting that mixed business with the refreshing fellowship of one's peers. This men's club quality was held up by anniversary spokesmen to contrast it with the level of public consciousness so broadly claimed for the Society in 1930. But the earlier picture has always had appeal for many in the Society, evoking an image of simpler times, good companionship, and an honest concern for technical excellence.

That sense of a golden age, when a man's word was his bond, also suggests a time of institutional innocence. Frederick Hutton claimed that there were no mechanisms for social exchange in 1880, and no periodical outlets for the detailed analysis of specialized technical problems. Furthermore, in his opinion neither of the existing national engineering organizations – the American Society of Civil Engineers (ASCE) or the American Institute of Mining Engineers (AIME) – lent themselves to the purposes of men devoted to machine design and construction, the generation of power, and the production processes of industry. What Hutton described, then, was a sort of vacuum; the field of mechanical engineering had grown in size and importance and the number of mechanical engineers had also increased, but no institution existed to satisfy the needs of that emerging profession.

The problem was not really one of institutional poverty, however. There were already a number of technical magazines, for instance, that served the community of mechanical engineers. The most important of the commercial publications were *American Machinist* (founded 1877), *Iron Age* (1859), and *Railway Age* (1870), although there were others. Among professional periodicals, the *Journal of the Franklin Institute* had been publishing the results of practice and research in mechanical engineering for over fifty years before

ASME's founding, while the transactions of civil and mining engineers carried articles written by mechanical engineers and read by them.

And mechanical engineers in 1880 were not without associations for social
purposes and the friendly communication of practical information from personal experience. The mechanics' institute movement that had flourished in
America in the years before the Civil War left technical libraries in many
cities that, as in San Francisco, often served as a focal point for people with
engineering interests. A number of these institutes also sponsored exhibitions
of industry, in places such as Cincinnati, which annually gathered technical
men together. Besides, there were established local engineering societies in
Chicago, Boston, Philadelphia, and St Louis, while the Cleveland Engineering Society was formed the same year as ASME. And, in fact, many of those
most active in the Society's founding were prominent in ASCE and AIME.

None of these other institutions was capable of projecting a national role for
mechanical engineers, however, and in the period around 1880 there were
powerful motives to organize the field along those broad lines. One source of
pressure grew out of the work mechanical engineers did. In some kinds of
industry – notably the construction of machine tools, steam engines, and
other machinery – it had become clear by the 1850s that economic success
depended upon advanced technical knowledge and that institutions were
needed to systematize the flow of information from research and experience
to practical application. Simultaneously, industrialists began to recognize a
parallel demand for the systematic analysis of production processes. William
Sellers' 1863 proposal for a standardized American screw thread was one of
the first efforts to extend this rationalization of manufacturing beyond the
level of regional practice, and it suggested a whole range of activities for a
national engineering society.[1]

Another pressure came from engineers themselves. As the American
economy expanded in the years following the Civil War, and as business
consolidation created increasingly larger firms, a growing number of managerial positions were created in the mechanical industries. The men in these
positions sought a distinctive career indentification and looked for it in
organizations with national standing and distinction. And that impulse was
not unique to salaried managers. Even those early leaders of the Society who
already enjoyed affluence and established position from the ownership of
large concerns also enjoyed the status attached to high office in a prestigious
professional society. In time, mechanical engineers would be anxious to

1 Alfred D. Chandler jr *The Visible Hand: The Managerial Revolution in American Business*
(Cambridge: Harvard University Press 1977) 240–81

define professionalism much more explicitly but from the outset ASME culti-
vated the image of an association of men whose talents and contributions
entitled them to a special place in American society.

The founding of ASME was also part of a larger process of institution building during the 1870s and 1880s. That period witnessed the widespread establishment of technical schools – Purdue, Stevens Institute, Case Institute,
and Cornell's Sibley College, to name several. It was also a time of explosive
growth in professional organizations. More and more people with specialized
knowledge formed professional societies and claimed professional status.
Both of the other national engineering societies belong to this period of institutional inventiveness. The American Society of Civil Engineers dates its
origins to 1852, but that first effort to found the society proved unsuccessful
and it was not until the early 1870s that the organization began to function
effectively as a national society of civil engineers. The American Institute of
Mining Engineers was formed in 1871, and, within the next decade or so,
such diverse associations as the American Chemical Society, the American
Neurological Association, the American Historical Association, the American Climatological Society, the American Economic Association, and the
American Institute of Electrical Engineers were established. These were but
a few of the groups seeking the same sort of specialized professional standing
that had impelled mechanical engineers to create a separate national society
of their own.

Finally, in the description of that array of factors that stimulated the establishment of ASME, one would have to return again to the role of the metropolis
where it happened. New York City contained the 'critical mass' necessary to
form a society with pretensions to national standing, as well as the leadership
such an enterprise demanded. It was not so much a matter of chance, then,
that the idea for a society grew out of conversations between engineers and
technical magazine editors and that New York–based engineers and technical
educators played major roles in its earliest years. Jackson Bailey, editor of
American Machinist, was only one of the journalists involved in the process.
James C. Bayles, editor of *Iron Age*, acted as secretary at the Stevens Institute meeting of 7 April 1880; Matthias Forney, editor of *Railway Age*, served
on the first Committee on Publication, as did William Wiley, of John R.
Wiley and Sons, who was a member of the Committee on Finance, too.

New York became the center of American engineering just as it had earlier
become the center of commerce, banking, and publishing. That dawning
awareness of being at the middle of things is revealed by the first steps
toward a club in the city to promote 'greater union among engineers.' In a
way that echoed ASME's beginnings, one of the preliminary meetings that led

to the formation of the Engineers' Club involved a technical journalist, Bayles; a technical educator, Professor Thomas Egleston of Columbia University; and a prominent mechanical engineer, J.F. Holloway, fourth president of ASME. Since AIME was without a building of its own at that time, ASME's founding also led to the first suggestion of a headquarters to house jointly all three national societies. It is worth pointing out that the primary objective of such an enterprise was to provide for 'the members in the city.' Once that was done, ASME President E.D. Leavitt claimed 'it will form a center for the engineering interests of the whole country.'[2]

The fact that there was a concentration of engineers in the city also meant that ASME's founders could use their experiences in other New York–based organizations to help in establishing the new society. Connections to the American Institute of Mining Engineers, for instance, proved especially important. Alexander Lyman Holley, acknowledged by all as the genius behind ASME, had served as AIME's president in 1875, while Robert H. Thurston was a vice president from 1878 to 1879, the year before becoming ASME's first president. There were a number of other links between the two organizations. Eckley B. Coxe served as president of both societies as did Robert W. Hunt. William Metcalf was president of AIME in 1881 and then for the following two years was a Vice President of ASME, while both Theodore N. Ely and William B. Cogswell, elected managers in ASME's first council, were also vice presidents of AIME. These personal links inevitably meant that ASME also adopted some of AIME's procedures. The process for election of officers, for instance, was taken directly from AIME, as were the categories used to define different grades of membership.

The American Society of Mechanical Engineers sprang to life, then, from a rich and varied institutional base, during a period of enormous vitality in the creation of national organizations for a myriad of special purposes, and at a time of great expansion in industries that depended on a high level of technology. These forces pulled men of varied backgrounds into the Society. Its early meetings reflected such disparate elements as the *conversazione* of European scientific societies, the mutual instruction of mechanics' institutes, and the applied physics of an emerging style of engineering science. Its membership mixed wealthy and powerful men in command of large enterprises employing hundreds of people with salaried superintendents whose managerial skills were called into existence by growing firms. These two groups constituted the majority of the membership, while the rest consisted mainly of professors of mechanical engineering and a few consulting engineers. As Monte Calvert has so well described, their differences in experience and expectation colored their views towards the Society's meetings, publications, and educational activities.[3] But they also had much in common.

Overarching all the differences, all the tensions inherent in the organization from the beginning, were some fundamental convictions about mechanical engineering and its economic and political relations. These assumptions were articulated in a remarkable fashion right at the beginning.

Robert Thurston, ASME's chief theoretician and first president, provided in his inaugural address a philosophy for the organization, a set of objectives, and a method for achieving them. Drawing his lessons directly from Herbert Spencer, 'that engineer who has become the greatest philosopher of our age,' Thurston outlined an approach for the profession that neatly integrated what he called 'scientifically correct conduct' with the kind of behavior enjoined by ethical and religious canons. It consisted, he said, of 'the care of self, of family, of friends, of fellow-citizens and of mankind' in a way that simultaneously protected individual interest and prevented injury to one's neighbor. According to Spencer, Thurston claimed,

perfect conduct ... will be attained when we shall have learned to fully protect and preserve self and family, and to individually attain length of life, perfect health and unalloyed happiness, while yet, in the same degree of completeness, promoting the same end as sought by each of our neighbors and by society at large.[4]

This vision of a world motivated by self-interest but served by mutual co-operation and aimed at mutual welfare was enormously appealing to Thurston's audience. In the first place, it justified their own ambitions for wealth and status. If a man's entrepreneurial talents created employment for others, he satisfied his own legitimate personal hopes in a way that gave others an opportunity to do the same, to the general benefit of society. The formulation provided the sanction of modern rationality to the biblical judgment that a man rightly prospered from the exercise of his talents.

Spencer's ideas also harmonized nicely with the concept of professionalism. The fact that the mechanical engineer was a professional man and that his knowledge was based on science, as the foundation was so often characterized, gave him a posture of objectivity and, as Burton Bledstein has

3 Monte Calvert *The Mechanical Engineer in America* (Baltimore: Johns Hopkins University Press 1967)
4 'President's Inaugural Address' ASME *Transactions* I (1880) 15

pointed out, great scope for his creative energies.[5] A profession at once conferred on its members democratic freedom of action and an elevated social status. It carried with it the notion of social responsibility and the implication that the professional labored to the larger welfare of the community as well as to his own account. And the image of mutual co-operation served equally well as a model for ASME itself; officers, members of Council, the Secretary and his staff, and the rank and file of the membership were all made equal partners in a noble enterprise, even if the profession also consisted of an elaborately graduated hierarchy, with ambitious men striving for its highest positions.

Finally, it was easy for mechanical engineers to believe that the ideal comforts and happiness Spencer described depended on a material abundance of the very sort their own kind of work created. Thurston reminded them that they directed the labors of nearly 3,000,000 working people, that they profitably employed $2,500,000,000 worth of capital, annually disbursed $1,000,000,000 in wages, consumed each year $3,000,000,000 worth of raw materials, brought into play motive energy equivalent to 3,000,000 horsepower, and that they used all these resources to create manufactured goods with a value each year of $5,000,000,000. Their skill and enterprise had given the United States a world-wide reputation for ingenious machinery, just as the country had become a haven for the industrious and ambitious of other lands. Touching on a theme ASME's fiftieth anniversary planners subsequently found equally appealing, Thurston claimed that the work of mechanical engineers had not only made America the most happy and prosperous country in the world, but had also created the essential qualities of modern civilization.

Thurston's argument was extremely important since it connected mechanical engineers to a fundamental theme in American culture – mankind's regeneration in the New World. He linked Europe's literary and philosophical traditions to an attitude that celebrated an abstract kind of intellectualism and despised 'the hard-working mechanic.' What America had taught the world instead was the value of the useful arts, the salubrious results when a society trained its citizens in 'mind and hand together.' That was the only sound basis, Thurston said, for an equitable distribution of the wealth required for happiness, for stable government, and for national prosperity. Modern civilization, 'reaching its highest development in this new world,' gave reality to the concepts of liberty, equality, and brotherhood. In a subtle and appealing fashion, Thurston thus wove together deeply embedded ele-

5 Burton Bledstein *The Culture of Professionalism* (New York: W.W. Norton 1976)

ments of the national ideology and themes that had profound meaning for engineers. Their own ambitions for a social and political status equivalent to what they saw as their major contributions to the country's development were made an essential element of America's mission in the world. Thurston made it seem as if they were all part of a crusade to give working men the opportunity to escape the centuries-old restrictions of social class and political subordination, even though his audience consisted almost entirely of men in quite comfortable circumstances who had never been in want. And the compelling part of it all was that Thurston grounded his propositions on the rightness of inductive philosophy. The 'principles' he elaborated to guide the newly established ASME thus also contained both a program and a method.

The inductive method, of course, flowed directly from the philosophy come to fruition in the scientific revolution of the seventeenth century. In its traditional form, it consisted of the collection and study of facts, the analysis of their relationships, and the comprehension of the natural laws that underlay those relations. Thurston gave that formulation a cast, however, that made it particularly appropriate for the new Society and its members. He made the collection of data, for example, a problem of information storage and retrieval. Facts should be registered 'in the most accurate possible manner, and so systematically and completely that they shall be readily and conveniently available, and in such shape that their values and mutual relations shall be most easily detected and quantitatively measured.' But even more, the professor, and future head of Cornell's Sibley College of Engineering, imagined a technical research establishment and posed as its first assignment to determine 'what path research shall take' and how money shall be found to support 'these self-sacrificing students of science, pure and applied.'

The answer to those questions was to be found in the Society Thurston described for his listeners. The most pressing demand of the future was 'to narrow the gulf which has separated men of business from men engaged in study, in experiment and in diffusing useful knowledge.' There was, he insisted, a community of interests between these two groups and it was the new organization's most crucial task to unite them. But Thurston meant a relation more complex than the romantic partnership of engineering and finance depicted in *Control*. He imagined two interrelated spheres of action. One consisted of 'men of the world' who were principally dedicated to the acquisition of wealth. The other was populated by 'men of science' – inventors, those whose practical observations might shed light on the improvement of techniques, men engaged in research, and the teachers of mechanical engineering. Men of science, in other words, were all those in any way concerned with the creation of knowledge or with its diffusion. They were the natural membership of ASME and the Society was created, in large part, to co-ordinate their efforts, identify the most promising lines of research, find the necessary funds, and then insure that the results were most effectively made usable.

These two spheres of action were dependent upon each other in certain important respects. On one hand, the prosperity of modern industry abso-lutely required the search for technical knowledge and its effective utilization while, on the other hand, research in applied science could not be pursued without financial support. Those were the two crucial points of connection, but the relation involved more than the exchange of money for services. There was an array of political activities that also fell within the area of their overlapping interests. In Thurston's mind, it would be appropriate for ASME to join with other professional societies in urging the United States Geologi-cal Survey to prosecute a thorough assessment of the country's mineral resources, or in conjunction with manufacturers' associations to press upon Congress the necessity of a government-funded program of materials test-ing. It was part of the Society's obligation to speak out in promotion of the useful arts, to make legislators aware of industry's educational needs or of the wisdom of tariff protection in the case of an emerging field of manufac-ture that needed it. Their interests, Thurston told his fellow engineers, had too often suffered from laws framed in ignorance or selfishness and he called upon them to act collectively to insure legislation favorable to their business concerns. The Society constituted a 'union of citizens having important interests confided to them,' he argued, and it was time to exercise their united strength in the promotion of those interests.

The American Society of Mechanical Engineers could hardly have found a more effective spokesman. Thurston used the democratic ideology of the mechanics' institute movement, Spencer's social Darwinism, and the tradi-tional notion of America's unique destiny among nations to argue for a new set of relations between capitalists and technical men and for a new kind of institution to embody those relations. In that light, ASME's founding can be seen as part of the infrastructure modern industrial capitalism created during the years after 1880 – as one of a set of institutions, along with technical schools and trade associations, fashioned to aid in the rationalization of large systems of production and distribution. Much as the Society emerged from the interplay of large historical forces, however, it would be a mistake to underesti-mate the role of men such as Thurston and Alexander Holley, whose vision of complex and technologically sophisticated industries depended upon the identification of an elite group of engineers made powerful by their command of a body of systematic knowledge. In the same way that Holley saw all the varied elements to the problem of translating the Bessemer process into American usage, Thurston perceived that the technological problems posed by the large-scale rationalization of the mechanical industries called for a different and more highly organized information system. At a time when most mechanical engineers were trained on the shop floor, he recognized the value of formal education in applied science, of engineering laboratories for systematic research into the materials and processes of modern industry, and of an agency that could integrate new knowledge into practical operations.

For all its emphasis on rationality, however, Thurston's conception of ASME was splendidly romantic, imagining as it did an equal partnership between men of the world and men of science (leaving aside the validity of the categories themselves) and a perfect congruence between private industry and public welfare. Those built-in strains would prove enduring problems for the Society, as would some of the organization's internal arrangements framed at the same time and in the same spirit. But Thurston's vision, of an association of professional men dedicated to America's technological pre-eminence and to a standard of material abundance for the country beyond any in the world, was as powerful as it was irresistible. His inaugural address was thus a marvellous performance, ideally suited to the occasion and full of insight into the needs of mechanical engineers and their employers. He had, as it were, conjured up a whole organization complete with membership, function, style and philosophy. It remained only to see how it would all work in practice.

A UNION OF MEN WITH IMPORTANT INTERESTS

The most immediate business of ASME was to work out the details of electing officers, admitting new members, conducting the Society's meetings, and handling the business that flowed from them. Models for all these activities were at hand in the practice of other societies and for the most part they were organized with relative ease. Only a little experience was needed, for example, to put into effect the practice of printing in advance the papers to be delivered at meetings, as the civil and mining engineers did, to facilitate their presentation and discussion. And at a time when all the members sat together in the same room to hear papers, it soon became apparent that a system was needed to prevent some speakers from talking too long. The rules to prevent this were not always consistently enforced, but these were not problems of any great moment.

And, at the beginning, the governance of the Society did not prove difficult. Alexander Holley had already outlined an approach to membership
requirements and to political authority. At the preliminary meeting of the
Society in New York on 16 February 1880, he described the ideal solution to
membership. It would combine the practical advantages of AIME's open
admission policy – a large membership, ample funds, and active connections
with interested capitalists – together with the esteem of elevated professional
status that ASCE's more restricted membership policy conferred. What he
proposed was a high degree of professional attainment for admission to full
membership and a category of 'Associates' for those 'men of the world' who
wished to join with engineers in pursuit of common goals.[6]

To carry out this program, Holley devised a two-tier admission process.
The Council, composed of the president, six vice presidents, nine managers,
and a treasurer, judged the fitness of each applicant and then presented all
those it accepted to a general vote of the membership. However, since the
actual wording of the membership requirements was left rather vague, the
Council had effective power in the matter. And it took that job seriously,
spending a great deal of time in the close scrutiny of applications for admission. In practice, the method was a bit different from what Holley had proposed, although the effect was the same. Demonstrated achievement was
still the basis for admission, but the Council had the latitude to determine
the kinds of personal success that best suited the Society's purposes.

Indeed, the Council rather quickly gathered power into its hands and that
was where the Society's projectors generally thought it should be. There was,
for instance, no challenge to the Council's filling its own ranks to make up
the loss of a seat through resignation, any more than there was to the president appointing the committee to nominate new officers. In the very early
years, before the Society had slipped into the comfort of established ways,
some members felt that the nominating committee ought to give the membership a choice of candidates rather than an official slate. The initial discussions about publications policy, over which the Council came to exercise
complete sway, also contain hints that not all members favored such an arbi-
trary style of proceeding. And at the third annual meeting in 1882, there
were objections raised to the Council's choice of season for the gathering and
to the way finances were handled. But these were not serious disputes and
the tenor of discussion makes it clear that the majority of those attending the
Society's winter and spring meetings favored a system in which a relatively

6   A.L. Holley 'The Field of Mechanical Engineering' ASME *Transactions* I (1880) 1–6

small number of men picked Council members, determined meeting locations, regulated admission, and played a large hand in the approval of papers for meetings as well as in their subsequent publication in the *Transactions*.

ASME's Council, after all, was made up of men who had achieved substantial success in life. It not only seemed proper to reward the leading men of the profession with the Society's highest office, but the wisest course as well since their undoubted abilities could only aid the organization's prosperity. The great majority of the Society's early presidents, for example, either owned and managed large industrial concerns or were the chief superintendents of such firms. Some, such as Eckley B. Coxe, were men of considerable inherited wealth and social position. Others – Henry R. Towne and Coleman Sellers, for instance – came from established families and broad cultural backgrounds; but even those early leaders of the Society without obvious family advantages, such as George Babcock, William Kent, and Joseph Holloway, also combined in their lives technical skill, business acumen, and a wide range of non-technical interests and experience. For the most part, these were men who knew machines and how to use them, who were as familiar with a balance sheet as they were accustomed to managing a workforce. They were models of career success and their own standing in the field of mechanical engineering helped confirm, for the membership generally, Thurston's claim that the Society was a union of men with important interests confided to their care.

That same sense of consequence gave a distinctly prosperous style to ASME meetings. Whether in Hartford or Cleveland, Chicago or Richmond, members had placed at their disposal chartered boats and waiting carriages, special trains and private cars. In Cleveland they were treated to a musical evening in Mark Hanna's own opera house; they were the Mayor's guests in Boston; Mr and Mrs George Westinghouse provided an orchid-filled ballroom and two orchestras for entertainment when the Society met in Washington. They were welcomed to these cities by governors and municipal officials and so fêted by local committees with excursions, collations, and dress balls that it finally became necessary for the Society to assume the costs of the meetings in an effort to regulate a ruinous competition of hospitality. The 'out-of-town' meetings were an important feature of ASME's early life. Until the twelfth annual meeting in 1891, when the Society settled into the practice of holding its winter meetings in New York, more than two-thirds of its spring and winter conventions took place in other cities. The Society's practice in these early years was influenced both by AIME, which moved about the country three times a year, and by the American Association for the Advancement of Science, which had long followed a peripatetic pattern of annual meeting. Gathering in cities other than New York also helped exercise the Society's claim of national status and undoubtedly served to increase membership, too.

In the first decade, attendance was never large. On average, about 85 members attended the spring meetings while the annual meetings attracted around 140. The percentage of membership attending meetings fell from 50 per cent at the first annual meeting to about 20 per cent at the tenth annual meeting, reflecting the Society's total membership growth, which during the same period went from 161 to slightly more than 1,000. To put it another way, the decrease in percentage of members attending also shows the relative decline of New York as the home of most members, even though more engineers from the city and its immediate vicinity attended the meetings than from any other geographic region.

The procedure normally followed at meetings further reflects the style of the organization in its early years. Members generally gathered in the evening of the first day for opening ceremonies which, after 1884 customarily included the retiring president's address. Then the meeting was usually adjourned for what the *Proceedings* often described as a '*conversazione*' and light supper. The following morning was devoted to Society business and for the next two or three days members heard papers, took excursions to nearby industrial establishments and educational institutions, engaged in discussion on practical questions called 'topical queries,' and were entertained by the local committee. The Society's meetings did not take hold in the immediate way official histories suggest, but after the third annual meeting, which seemed the low point of the organization's early fortunes, subsequent conventions waxed ever more affluent and congenial. Growing numbers of wives attended programs devised especially for them, and in time the meetings actually became reunions in the way the founders had hoped. The size of meetings also encouraged that tendency. There were no simultaneous sessions in the nineteenth century; everyone sat together and both the professional and social elements of the program emphasized personal interchange. As meetings took on a comfortable and familiar quality, Joseph Holloway's easy affability seemed more in order than the self-conscious formality Thurston had imposed by referring to members always as 'the gentlemen.'

The secure sense of accomplishment and standing that members derived from their meetings was nowhere better expressed than in the European excursion of American engineering societies in 1889. The idea for it, once again, had come from Alexander Holley's fertile imagination. His own frequent trips to Europe on behalf of American Bessemer steel interests had not only led to friendships with British and Continental engineers, but had raised in his mind the concept of an international technical fraternity. By 1881 that notion had taken the form of a joint committee of civil, mining, and mechanical engineers whose charge was to plan for a visit of foreign engineers at the invitation of the American societies. Since Holley was shortly to go abroad, it was left to him and another member of the joint committee to sound out the Europeans, particularly members of the British Iron and Steel Institute. But the project foundered; Holley fell ill and the plan itself was undermined by some unidentified sort of competitiveness. As ASME's Council obliquely reported, 'the attitude taken by members of the profession over there was not precisely what was desired by our members.'[7] That feeling of rebuff echoed at the memorial meeting of the Society in April 1882 to mourn Holley's death. In a tribute to his achievements, James C. Bayles remarked that 'petty jealousies' had militated against the British Iron and Steel Institute awarding Holley its Bessemer Medal.[8] A lingering sensitivity between the organizations also prevented an attempt to arrange a New York meeting of the institute in 1884 when many of its members were to be in Montreal for the British Association for the Advancement of Science meeting.

Counterbalancing these impediments to institutional co-operation were some practical realities. The normal pursuit of business sent Americans abroad, just as it drew Europeans to these shores, and in particular the international industrial exhibitions so popular in the nineteenth century focused attention on the exchange of ideas. Thus it was that the Paris Exhibition of 1889 provided the opportunity for the kind of meeting of technical men Holley had imagined. The president of Britain's Institute of Mechanical Engineers, having learned privately from ASME Treasurer William Wiley that American engineers would be attending the exhibition, formally invited the Society to meet in London first, assuring a warm welcome from the Iron and Steel Institute and from the other engineering societies. From that point, the affair quickly took on an ambitious style. ASME chartered a steamship of the Inman Line, the *City of Richmond*, for the voyage and had it refitted so that all cabins were converted to first-class accomodation. The trip itself, the Society's Secretary Frederick Hutton recalled, turned into a 'large yachting party,' with games and sports en route.[9] But the engineers also took seriously

7 ASME *Transactions* III (1882) 6
8 Ibid 46
9 Frederick R. Hutton *A History of the American Society of Mechanical Engineers* (New York: ASME 1915) 230

the formal exchange of courtesies and appointed a shipboard committee to prepare for the festivities that awaited them.

The excursion had in the meantime swelled to include ASCE, AIME, and a few representatives of AIEE, while in Britain the Institution of Civil Engineers had taken the overall responsibility of organizing hospitality for the visiting Americans. Even for an era practiced in ornate ceremonials, the reception of the visiting engineers was remarkable. The heads of Liverpool's largest shipping interests met the *City of Richmond* even before it docked to tender an official welcome and convey the party through customs and to their accomodations. In the two weeks that followed, the American engineers were treated to an absolutely dazzling round of luncheons, evening banquets, receptions, excursions, inspections, and open houses where their hosts included such prominent British men of engineering, science, and industry as Sir Henry Bessemer, Sir Frederick Bramwell, Professor Unwin and Professor Tyndall, Lord Brassey of railway construction fame, and Laird the shipbuilder. They received free rail transportation throughout their stay in England and were shown the outstanding engineering points of interest; they were fêted by the Lord Mayor in London's Guildhall and were shown the Queen's private apartments at Windsor Castle. And that was not the end of it, by any means. They were taken to Dover 'on a magnificent special train,' crossed the Channel in bright sparkling weather on a steamer chartered for them, and were met at Calais by dignitaries of the French Society of Civil Engineers, who were waiting with a special train of their own to carry the party to Paris. There the whole grand process, including an extended trip on to Germany for those who wished, was repeated. And when they returned to New York, Joseph Holloway welcomed them back with a 'handsome dinner' at the Engineer's Club.

This triumphal month-long tour of the Old World touched ASME and its members in a variety of ways. At a personal level, for instance, the trip helped give Frederick Hutton a sense of importance he never lost and he also discovered, as subsequent secretaries did, too, the charms of moving in international engineering circles. European travel was not a novelty to many of the Society's leaders. Ambrose Swasey and George Babcock regularly vacationed in Europe, John Sweet had trained there as a young man, and there were many others whose business interests required foreign visits. Eckley Coxe had studied in Paris and Freiburg while Henry R. Towne, chairman of the joint committee representing the American engineers, was fluent in French from his Continental travels. But in 1889 they went abroad as delegates of the American engineering profession. They had wealth and leisure enough to take such a trip and they were men full of confidence in the country's industrial development and in their own role in that process. The only possible doubt was what status they would have amidst long established European institutions. However, in England, especially, they discovered a surprisingly strong sense of mutual purpose that flowed from an awareness of the power of modern industry and a feeling that engineers might form a new kind of aristocracy. The American minister caught some of that spirit in his Guildhall banquet speech when he told an appreciative audience that engineers were doing more 'to bring about the brotherhood of man' than any other agency in the world.[10] From the official account of the trip it is clear that ASME's members returned sure that they were equal partners in a professional movement of international dimensions.

Strong as those feelings were, and significant as they were to prove for the development of international standards, what most of those who went abroad on the excursion of '89 probably remembered best was its sociability – the grand treatment they had received and the special feelings of comradeship within the group that had grown out of their travel together. That was another of the important results of the voyage. It fixed a certain style on that part of the Society's activities usually described as 'the social element.' Holley had claimed that the organization's social role was just as important as its technical pursuits. The trip to Europe was that concept worked out in practice, and just as the Council's authority in the Society tended to concentrate the management of its affairs in the hands of a few leading members, this style of social activity was more easily dominated by those of wealth and power.

There is no evidence to suggest, however, that during these years very many of the Society's members objected to that form of social expression. In fact, the Council proposed holding the 1892 spring meeting in San Francisco, which would obviously be costly and time-consuming to attend, and very carefully canvassed the membership for reactions to the idea. The Council argued that a meeting in that part of the country would establish beyond doubt the Society's 'standing as a national organization,' that it would generate members west of the Mississipi where few existed, and that it would have the same public relations benefits as the European trip. Within the Council's debates, the only real drawback seen was in setting a meeting 'where only a wealthy and leisured minority could attend it.'[11] The circular letter to the membership did not mention the problem in that explicit fashion, but rather emphasized the good effect of the voyage to Europe and the similar potential

10   Ibid 235
11   ASME Council Minutes 16 Jan. 1891

in this proposed jaunt to the Pacific coast, which in like style would involve a month-long excursion on a specially chartered train, with side-trips to points of particular scenic interest. About half of ASME's 1443 members answered the circular letter and only 10 per cent of them opposed the idea – almost exactly the same number, 75, that actually took the trip. In Secretary Hutton's opinion, the group was 'just the right size ... everyone knew every one else.'[12]

THE SEARCH FOR POLITICAL POWER

Even though he died early in 1882, Alexander Holley's name continued to appear in ASME's *Transactions* for years afterwards. It was partly because he had had such an acute sense of institutional politics; he knew the best way to get things done, and that talent was often recalled in business meetings when matters of policy were being discussed. In an even more compelling manner, Holley reminded the Society's members of the high ambitions they had started with. Robert Thurston made that point as he reminisced about his own relationship to Holley at the memorial service for him. They were both Brown University alumni, had worked together, and shared similar tastes. But what really bound them in rare and close friendship, he said was 'a common and more intense interest' in 'The Scientific Method of Advancement of Science.'[13] What Thurston meant by that expression and what he and Holley dreamed of was a scientific kind of engineering that systematically organized mechanical knowledge and discovered its most efficient application to practical purposes. That passionate concern they both felt for the promotion of mechanical engineering – as a body of knowledge, as a profession, and as a great engine for America's industrial development – imagined elaborate programs of technical research, a highly structured system of technical education, government support for both, and perhaps even some supreme scientific council or academy to co-ordinate these activities and advise the government at the highest levels of policy making. This splendid vision, exactly like that of American scientists in the years before the Civil War, continued to inspire the Society as its most active members sought to translate that early idealism into practice.

One of the very first questions posed by Holley's program involved politics: what kind of political stance should the Society take in order to achieve its aims? Thurston had argued for concerted action in his inaugural address,

12 ASME *Transactions* XIII (1892) 703
13 Ibid III (1882) 30

but, even before that, Holley had confronted the issue by his leadership of a movement to persuade the government to continue its support of research into the strength of iron, steel, and other construction materials. The Iron and Steel Board had been established by Congress for those purposes in 1875. Thurston was its secretary, Holley one of its members, and William Kent, later prominent in ASME, also worked on the project. Despite delays in acquiring test apparatus and a quarrel with the Army Ordnance Bureau, which had an interest in strength-of-materials testing, the investigation achieved some notable results before Congress cut off its funds in 1879. Holley had been active, through his connections in ASCE, AIME, and the Iron and Steel Institute, to marshal support for the restoration of the board's budget. After ASME was founded it, too, was enlisted in the cause and became the leading element in it following Holley's death.

Those most involved in the campaign saw it at first as a problem in education and in political reform at the grass roots level. The first task was to help Congress and the general public understand the importance of the information that the research provided. Steam-boiler explosions and such disasters as the collapse of the Pemberton Mill in Lawrence, Massachusetts, from the failure of its iron supporting columns, helped raise interest in the matter, but from an engineering point of view it was even more important to know with certainty the qualities of material that manufacturers daily supplied for construction purposes. Furthermore, according to William Metcalf, a Pittsburgh engineer connected to the iron industry, the amount of information and level of detail required made any such investigation absolutely beyond the ability of a manufacturer or a technical association to finance. Public funding was the only answer and, he said, evoking a picture of fruitful co-operation between government and industry, it was also the most appropriate solution since the results would so obviously benefit national prosperity.[14]

In a confident mood, the Society held a special session, at its spring meeting in Philadelphia in 1882, on the need for strength-of-materials testing. At a moment when the organization could scarcely bear the cost, the proceedings of that session, together with supporting letters from concerned engineers, were printed in a pamphlet for wide distribution – and then reprinted and mailed out again two years later. Besides that kind of educational effort, many ASME members connnected the issue to the more general question of political reform. As William Kent claimed, if good men could be induced to run in the primaries against candidates put up by 'the ring,' there was a chance for favorable Congressional action. But how to

14 Ibid 121

'engineer' the present members of that body, he confessed, was beyond his knowledge.[15]

That mixture of optimism and despair was characteristic of most of the political causes the Society took up in its early years. In the case of government support for strength-of-materials research, a great deal of effort was spent in an attempt to identify the levers of political power in Washington and how to use them. But, after four years of labor and the use of 'every honorable means within their knowledge,' the committee had still failed to influence Congress and its chairman was forced to ask that it be discharged from any further responsibility.[16]

The hope for a new political order, in which the rational perspective of engineers and other men of affairs might predominate, seemed to fade with each encounter. In 1885, for example, Oberlin Smith, a Bridgeton, New Jersey, machine works president, called for ASME to take action toward a reform of the Patent Office. The affairs of that department of government seemed proof of Thurston's inaugural charge that it was time for the country's industrial interests to take action on their own behalf. Not only was the Patent Office crowded half out of its own building by other bureaus, but it was also six to eight months behind in its work for lack of examiners. And staff salaries were so low it was difficult to attract competent people. But all the fees inventors had paid, amounting at that time to $3,000,000, had simply accumulated in the treasury, Smith claimed, rather than having been rightfully employed in meeting the needs of the technical community.

It was much less clear, however, what the Society could actually do about the situation. Smith himself had mixed feelings about politicians. As he put it in a joke that still carried the sentiments of many engineers, there should be two Congresses, 'one to attend to politics and one to attend to business. The only other way I know of is to kill all the present Congressmen and put members of the engineering societies in their places.'[17] Professor Egleston, of Columbia University, thought the best hope for any kind of political success was 'to increase the average intelligence of the ordinary Congressman,' while others saw the problem as one of civil service reform. In any case, a petition finally seemed the only thing the Society could do.

There were other causes that stirred ASME to some kind of political action during the last years of the nineteenth century, but not in a way that caught

15 Ibid 108
16 Ibid VIII (1887) 18
17 Ibid VI (1885) 18

the imagination or crystallized sentiment as in the early years. In practice, that part of the Holley/Thurston vision for the Society proved unworkable. It seemed impossible to connect matters of technical importance with issues that stimulated interest in Congress. But it was more than just a problem of naïvety. During that period Congress was ill-disposed to support any kind of scientific or technical research; and engineers did not function well in the political arena. Politics seemed a corrupt business to most of them, or certainly, at least, to those who played leading roles in engineering societies. More attracted to a professional self-image of objectivity and rationality, they characterized politics as an appeal to emotion and vulgar instincts. But, unlike many professionals of that period who sought to use their special skills to make government more effective, ASME's early experience with politicians left its members, as J.F. Holloway flatly announced, 'pretty well disgusted.'[18]

The idea of a national academy of engineering suggested a much more elevated avenue for political influence. Like so many of Alexander Holley's conceptions, it proved very appealing to members of the Society. The plan had not been well developed before he died, but in a general way Holley had sketched out an academy of engineering made up of the best men from each of the societies, drawn into a 'bond of sympathy.' He did not elaborate the program of such an academy, but just as he had imagined international connections between engineers beneficial he thought this union of American engineers of different perspectives could not help but be productive of much good. In any event, Holley's idea stimulated talk on the subject, and at the Hartford meeting of the Society in the spring of 1881, Oberlin Smith argued that some sort of agency was needed to act as a 'great National University of Science.' His concern was less with engineering society co-operation than with technical research and with a range of activities that centered on information retrieval and on standardization of the kind exemplified by the Franklin Institute's screw thread.[19] Without any influence in Congress, engineers could hardly expect the government to recognize 'the magnificent industrial economy' of supporting such an educational venture; so Smith suggested a 'central council' of machine makers and users, who out of their own pecuniary interest in the benefits of such a scheme ought to provide the funds for it. But while he argued for an industrially funded research establishment, his model was something half-way between *Engineering Index* and the National Bureau of Standards, neither of which existed at the time.

18   Ibid 24
19   Ibid II (1881) 56

In the discussion that followed, Thurston connected Smith's idea with the general problem of support for technical research. What one might have expected from history, he implied, was that the Franklin Institute's federally supported research on the strength of materials in the 1830's and Walter Johnson's subsequent investigation for the Navy Department into the properties of American coal, should have been followed by more extended research support from the federal government. But Congressional appropriations in 1872 to discover the causes of steam-boiler explosions or in 1875 for the short-lived Iron and Steel Board were either misused or terminated prematurely. And not much could be expected from the kind of research most manufacturers carried on, Thurston argued, since it was usually restricted to a limited set of problems. His observations brought the discussion back to where Smith had started it: there was a great need for technical research, a compelling demand for standardization, and some new kind of institution was required to solve these problems.

William Kent, Thurston's protégé, used Holley's early idea to frame the classic exposition of an American academy of engineering. In an 1886 address to the American Association for the Advancement of Science he outlined its features. It would be an aristocracy of engineers, intellectually important and politically powerful. The academy would have a large New York headquarters with library and research facilities, a museum and meeting rooms. It would conduct the research of the former Iron and Steel Board as well as other useful investigations, organize and oversee all the government's public works projects, maintain a technical college, and in other ways insure America's technological pre-eminence. Exactly like Alexander Dallas Bache's original idea for a national academy of science, Kent's proposal would have given engineers political authority and control over public funds without any form of Congressional oversight. It differed only from the scientist's academy in its election procedures. There would be 'no council passing upon nominations in secret conclave,' Kent said, inadvertently describing precisely the admissions system ASME then used.[20]

It was obviously appealing to imagine an academy of engineers as prestigious as the National Academy of Sciences, particularly since engineers were perfectly aware of the condescension their work often received. The idea of some kind of central organization that might concentrate their intellectual authority and political influence into a single body was also an idea that engineers continued to find attractive. But the fate of Henry Towne's effort to

20 William Kent 'Proposal for an American Academy of Engineering' *Van Nostrand's Engineering Magazine* XXXV (1886) 277–80

establish a unified group reveals other forces simultaneously at work. At the tenth annual meeting in 1889, Towne proposed that the Society appoint a committee of three to represent ASME in a conference with delegates from the other societies to explore the subject of 'a national organization of American engineers.'[21] The idea for it, he said, had arisen out of the experience of those who had gone on the excursion to Europe the previous summer. The British Institution of Civil Engineers, because of its claim that its title merely distinguished between civilian and military engineers, had been able to act on behalf of all the kingdom's engineers and from that position of concerted strength and influence mounted the lavish reception accorded to visiting American engineers. Towne's mind was much on the forthcoming World Engineering Congress, to be held in conjunction with Chicago's Columbian Exposition, and his proposal aimed at some means to return the same level of hospitality when foreign engineers would be coming to the United States on that occasion.

That objective was enough like earlier ideas of an academy or union of engineering societies to remind ASME members of Holley's dream. But in the ten years since the Society's formation its membership had grown to equal that of the well-established ASCE and there were many who saw unity as a threat to that sturdy independence. Joseph Holloway spoke for them when he said he did not want to see the Society 'overshadowed by any other.' Kent, who had a direct way of talking, pointed out a related issue that most of his colleagues usually dealt with in a less candid fashion. The civil engineers, he claimed, were trying 'to enlarge themselves by annexing the various local societies in Cleveland, St. Louis and other places.'[22]

In fact, all the national societies were struggling to make good their claim to country-wide representation, and cities with strong local engineers' clubs presented tempting congregations for conversion. These local groups were also a potential mechanism for connecting distant members to the New York headquarters, a problem none of the national societies had yet solved. Paradoxically, in a situation that looked like sheep chasing wolves, the city engineering societies that had taken root so solidly in the late nineteenth century started a unity movement of their own in 1885 and that, too, disrupted the hopes of men such as Towne, Kent, and Oberlin Smith. Furthermore, a small but outspoken minority in ASME objected to the academy concept on the grounds that it was 'an un-American idea,' tending to the creation of a

21 ASME *Transactions* XI (1890) 32
22 Ibid 41

class system within engineering.[23] And, finally, Towne's proposal did not meet with any success in the councils of AIME or ASCE. Both societies decided it was 'inexpedient' to pursue the matter any further.

The vision of an academy of engineering captured the most rarefied hopes of engineers for political power and intellectual status. It would prove difficult enough, in fact, to find consensus on many issues within ASME, let alone among the other societies, too. But the hope of a profession joined in some kind of academy or unified organization kept recurring, sometimes in other forms. Some mechanical engineers came to the conviction, alternatively, that the authority and political influence they sought might actually be best directed at the level of municipal politics. In time, too, the Society developed an array of prizes and awards to identify the sort of distinguished career an academy membership would signalize. Discussion of the unity issue suggested other possibilities. One of the most concrete proposals came from R.W. Pope, secretary of the American Institute of Electrical Engineers, who in the debate over Towne's idea mentioned his thought of a 'society of societies,' consisting of the officers of the different organizations meeting periodically to work out routine matters of mutual interest, a notion that implied the possibility of the secretaries of engineering societies doing the same thing. And in a way that could have brought a potential problem closer to resolution, talk of unity revealed that many engineers found quite a satisfactory professional affiliation in their local engineers' clubs.

Finding harmonious and effective means for inter-society co-operation was a difficult business under the best of circumstances, as subsequent generations of ASME members would learn. But curiously enough, of all the ingredients that combined to defeat Henry Towne's 1889 proposal, the Society's own vigor was probably the most important. Most members at that meeting could recall the very first difficult years of the organization and the New York meeting in the fall of 1882 that seemed the nadir of the Society's fortunes. But the Cleveland meeting the following spring, marked by Joseph Holloway's spirited hospitality, also stood out as 'a new departure for the Society,' and nothing had appeared to check its progress since. The number of members increased and more people attended meetings; 250, a record number, registered for the tenth annual meeting. Not only had ASME come to equal longer established societies in membership, but its financial prospects had also so improved that in 1891 it bought the New York Academy of Medicine's building on West 31st Street for its headquarters. But more than all these evidences of prosperity, what contributed

23   Ibid 601

most to the growing confidence of the mechanical engineers, and the concern that their society might be diminished in any unified organization, was the feeling that they were crucial elements in an unfolding historical process of revolutionary importance. Among all the engineering societies, they saw themselves as the chief agents of American industrial power. Holley and Thurston had encouraged them to that view, but in an even more compelling development, as they came to realize the enormous value to industrial productivity of standardized practices – whether in manufacturing or management – they saw an expanding professional role of fundamental importance. It became clear the Society could have enormous impact in these direct and practical ways. George Babcock, one of the founders of Babcock and Wilcox and Society President in 1887, best illustrated that kind of ambition when he proclaimed it ASME's mission to strive for the day 'when every force in nature and every created thing shall be subject to the control of man.'[24]

STANDARDIZATION AND THE EMERGENCE OF TECHNICAL PURPOSE

The first clear awareness of the economic potential in systematic technical knowledge had come to Philadelphia machine builders in the 1830s. In just the sort of co-operation between 'men of science' and 'men of the world' Thurston had said ASME was founded to bring about, a program of technical research and publication at the Franklin Institute joined a small group of young scientists looking for professional careers and a number of machine builders who were at the beginning of their business lives. Their investigations – into the most effective use of water-power, the causes of steam-boiler explosions, and the strength of materials – became models of technical research, while the resulting reports helped establish the publication of original inquiry as a criterion of engineering professionalism.

Out of that experience, entrepreneurs such as Samuel Vaughan Merrick and William Sellers made Philadelphia the center of American machine building in the years before the Civil War. The outstanding characteristic of their work was its 'rationality.' Foreign engineers praised Merrick's Philadelphia gas works for the uniform gauge of all its fittings, making it cheap and easy to repair the system. Sellers concentrated his mechanical skills on the construction of highly functional machine tools explicitly designed to increase output without an increase in labor costs, an approach he later brought to the manufacture of standardized bridge parts. And it was Sellers who first pro-

24   Ibid IX (1888) 37

posed an American standard screw thread system in a paper he presented at the Franklin Institute in 1864.

In a number of ways the Philadelphia organization served as a transitional stage in the professionalization of mechanical engineering and many men subsequently active in the formation of ASME had been involved in one way or another with the Franklin Institute. Thurston, for instance, published his first technical papers in the *Journal of the Franklin Institute*. Its editor, Henry Morton, learned of his work in that way and when he later became president of Stevens Institute appointed Thurston to the faculty. There were others who also published in the *Journal* to advance their careers, but the Franklin Institute ultimately proved an unsatisfactory vehicle for engineering professionalism because of its general character in an era of increasing specialization and because the machine building industry outgrew Philadelphia. Its role in generating an American screw thread standard, however, pointed to an important range of activities for a national organization of mechanical engineers.

ASME took up these kinds of concerns right from the beginning. Standardization provided an effective means of processing technical information and of integrating it into industrial practice. The simplest kind of standard might involve little more than an industrial census and some minor compromises to achieve uniformity. At the other extreme, the development of a standard could be an intellectually satisfying form of problem-solving that required elaborate research programs to create new knowledge. In any case, however, standards were aimed at systematic understanding and the rationalization of practice. Engineers usually found purposeful non-standard usage offensive to their sense of order and in their zeal for a particular point of view sometimes connected standards with national or cultural values, as in the long and noisy fight over the metric system. But William Sellers was at pains to point out, in arguing for the adoption of his own system of screw threads, that standards most of all involved economic considerations.

The society dealt with just that issue at its very first annual meeting. George R. Stetson, of Morse Twist Drill Company, read a paper entitled 'Standard Sizes of Screw Threads,' and Coleman Sellers, a partner with his cousin in William Sellers and Company, presented a paper on 'The Metric System – Is It Wise to Introduce it into Our Machine Shops?' Stetson's paper pointed out how complex a matter it was to secure an effective standard in screw thread practice. For instance, even if the shop-floor machinist understood the Sellers system, and according to Stetson there was much evidence to the contrary, the manufacturers of taps, dies, and gauges did not themselves supply standard tools. The Sellers, or Franklin Institute standard, as engineers often called it, was 'calculated to meet the requirement of the country,' Stetson said, and stood the best chance of adoption. But an effective level of compliance called for education, proper gauging tools, and their vigorous application. Only a 'central organization' such as ASME, he suggested, could bring that about.[25]

Coleman Sellers' paper also urged the Society to take action in the field of standards. A fervent opponent of the metric system, Sellers claimed that his firm's experience proved it was a less effective system to employ and that the change-over costs would be enormous. Instead, it was the particular talent of Americans to discard unsuitable European practices, he said, and to strike out in fruitful new directions. Thus, 'just as we have seen fit to drop the letter *U* from some of our words,' Sellers argued, the Franklin Institute standard boldly addressed deficiencies in Whitworth's system. That kind of progressive approach to technical problems had its own pay-off: 'He can command the markets of the world who can make the best machinery at the least cost.'[26]

Those were popular sentiments within ASME. 'I do not comprehend it at all,' Henry Worthington said of the metric system, and he went on to remark, 'I think I can safely say that I never met a single representative of our plain, practical profession who advocated it out and out.'[27] Not all of Sellers' listeners approached the subject with the same vehemence he felt for it, or came at with Worthington's directness. But the efforts of the metric system's advocates to have its use made compulsory seemed a sufficient threat to call for a reaction. Worthington, again, put it in the most straightforward way: 'For the first time in my life, I stand before a meeting that is capable of giving expression to the opinions of the mechanical engineering ability of the country, and I am anxious to see the development of their power in useful directions as soon as it can reasonably be made.' In his opinion, it was the Society's obligation to 'take a commanding position' in such affairs and he proposed a two-part resolution. The first was that the Society endorses Sellers' views on the metric system and the second that ASME 'deprecates any legislation tending to make its introduction obligatory into our industrial establishments.'[28]

No one disagreed with his propositions. Indeed, to make the point as strongly as possible, those at the meeting determined to have the resolutions

25 Ibid 1 (1880) 3
26 Ibid 15
27 Ibid 24
28 Ibid 26

cast in the form of a ballot to be sent to all the Society's members. At that moment, there were just under 200 members, of whom 135 returned their ballots, voting four to one in favor of the resolution. The vote itself is not surprising. What makes the event remarkable is that for years afterwards a few prominent members argued that the Society's settled policy was to take no position on standards.

There were in fact conflicting ideas within the Society on the subject. Some members felt the demand to systematize industrial practice by pre-scribed codes of an authoritative stamp. At the same time, the sort of profes-sionalism Thurston and Kent advocated would have the Society chary of tying itself to recommendations that could take on an essentially commercial quality. There was also a strong sense that standards should reflect the dis-tilled experience of the market-place, but against that, the awareness, too, that codified procedures were soon outmoded by technical advance. In other words, ASME's relation to standards was complex – not accurately indicated by the metric system controversy or by simple categories of opinion within the Society – and it changed over time.

The question of policy came up initially as a result of Pratt and Whitney's effort to produce a standard thread gauge, an instrument to solve just the problem George Stetson had described. George M. Bond, of Pratt and Whitney, had outlined, at the Hartford meeting in the spring of 1881, the method the company planned to use, and in the following year presented a paper on the successful results of that investigation, inviting ASME to inspect the process, in the interest of 'an impartial verdict as to its accuracy and practicability.'[29] The extraordinary precision of the work and the expense to which the company had gone to achieve it immediately excited the admiration of those who heard Bond's paper. 'Undoubtedly,' Professor S.W. Robinson of Ohio State University claimed, 'this is a thing which the Society will be willing and anxious to endorse as a standard of the country.'[30] William Kent pointed out, however, that AIME had a rule 'never to endorse anything,' and he proposed instead the appointment of a committee of experts whose find-ings would as a matter of routine be reported and published.[31]

The debate does not reveal that Kent or anyone else at the meeting saw a contradiction between a mail ballot on the metric system and the proposition that the Society should not endorse anything. The metric system was not

29 Ibid III (1882) 128
30 Ibid 129
31 Ibid

employed anywhere in American industry for items of domestic use, it enjoyed no real support in the Society at that moment, and its imposition promised only expense and inconvenience. Pratt and Whitney's 'Comparator,' however, even though a result of industrial research on a fundamental engineering problem, was still a device from which the company expected some profit.

If it was easy to accept the principle that the Society should not endorse a commercial product, what it should do with a standard generated by one of its own committees was yet to be determined. However, the eagerness with which its members assumed the Society would play a paramount role in setting standards meant that the issue would soon be raised. A standard for screw threads only started the rush. Oberlin Smith thought the Society should standardize shop drawing symbols, wire gauge, pulleys and line shafting, machine screws, nut bevels, key-seats, drawing boards, and mechanical dictionaries – and he meant merely to suggest the range of concerns appropriate for the organization to consider. The Franklin Institute was just starting a campaign to establish national standards for gear teeth, he said, and 'if a local society can accomplish so much, why should we not go further and give this and other matters the prestige of a national movement?'[32] What Smith ultimately wanted to see was 'a national bureau of information and standards,' since to him information retrieval and standardization were closely related, and he proposed the appointment of a committee to study the feasibility of that idea. But E.D. Leavitt, the Society's President in 1883, suggested the matter be referred to the committee just appointed to report on Pratt and Whitney's standard thread gauge. This move suddenly and substantially broadened the mandate of the Committee on Standards and Gauges, and seemed to commit ASME to an active role in a vitally important sector of mechanical engineering. Frederick Hutton was eager to have the Society also determine a standard for rating steam-boiler capability, and observed 'it is part of our duty, no doubt, to establish gauges and standards.'[33]

In the drive to rationalize American industry that began to gather force in the last quarter of the nineteenth century, standardization was to the engineer what administration was to the manager. Within the technologically complex mechanical industries, especially, the creation of standard parts and uniform practices gave the engineer control over anomaly. Standards yielded predictability and regularity in the production processes as well as greater power over the work-force and the work-place. In that sense, Frederick W.

32   Ibid 19
33   Ibid 20

Taylor's vision of 'scientific management procedures' was nothing more than the logical extension of a process that begun with screw threads.

Inevitably, the methods for developing standards became organized, too. Oberlin Smith's proposal for a bureau of information and standards reflected only the enthusiastic beginnings of ASME's long career of formulating codes and standards. By itself, Smith's idea had little to recommend it, since if nothing else it ran counter to the way the standards movement was developing in America. Without any applicable tradition of legislated standards (the Navy Department adopted Sellers' screw thread simply because it seemed the system most likely to be used by private industry and it was the only governmental department to take any action at all) it fell to voluntary agencies to create them. Organizations such as the Railway Master Mechanics Association and the Master Car Builders' Association, another railroad group, were among the first in the field, but standardization soon became the chief *raison d'être* for a host of institutions in the late nineteenth century. In a multi-dimensional institutional structure that grew increasingly complex with the years, standards were promulgated by trade associations, technical societies, joint committees, and finally, by umbrella organizations that systematized and integrated a widespread interest in standardized procedures.

ASME's own history of standards also became more elaborate in time, but it began with the sort of practical concerns of men who worked with steam-engines, boilers, and pumps. William Kent, who proposed that the Society devise 'a standard set of boiler testing rules,' stood directly in the Holley/Thurston tradition in ASME. He had been a student of Thurston and had worked as his assistant in the strength-of-materials investigation conducted at Stevens Institute for the Iron and Steel Board. In the same way that inadequate knowledge of basic construction materials inhibited their systematic use, the most effective exploitation of steam-power required that boiler capability be described in terms that both manufacturers and users understood. One of the problems that prevented the straightforward exchange of information between those two groups was that market-place competition stimulated extravagent claims for boiler performance, particularly in regard to the amount of fuel required to generate a given amount of steam. But what complicated the matter even further, to use Kent's words, was that 'every engineer who makes a boiler test makes a rule for himself, which may be varied from time to time to suit the convenience or interests of the part for whom the test is made.'[34] To sharpen the message, Kent described his own

34 Ibid v (1884) 260

method of conducting boiler tests, and the ensuing discussion, in just the way he had hoped, made his point that the differences in techniques between them called out for a standard set of procedures.

One of the interesting things about Kent's idea, which resulted in the formation of a committee charged to study the subject and report on it, was that it proposed to do exactly what Frederick Hutton had so unsuccessfully suggested the year before. But Hutton had proposed the Society establish a standard for boiler horsepower, whereas Kent argued for the necessity of uniform test procedures. One proposal would have committed ASME in a matter that was frequently the source of litigation, while the other enjoyed the posture of objectivity – even if the determination of the boiler's ability to generate power was what tests were always about anyhow.

Kent's report emphasized practical utility rather than scientific precision. He had aimed, he said, at a 'code for daily use by the practising engineer.'[35] Thus, for example, the Committee retained the unit of boiler power used by judges at the 1876 Centennial Exhibition to compare competing boilers on the grounds that it was a familiar standard and adequately represented 'good average practice.' Among others, Professor W.P. Trowbridge of Columbia University objected to Kent's approach, claiming that the lack of exactness detracted from the standard's authority. Clearly, that was one of the balances to be struck. In the past, boiler tests had often been carried out under less than neutral circumstances and for boiler makers, their customers, and the engineering profession the situation was awkward. Some kind of order was needed, but it had to come in a form that all sides were likely to find acceptable. Kent's code was nicely calculated to do just that and most of those who participated in the discussion supported it warmly. The boiler industry was continually disturbed by legal action arising from the disparity between claims and performance but, perhaps even more destructive to the orderly application of steam power, the engineering community had been without the means to deal with the problem in mutually agreed-upon language.

In fact, the report seemed so important the Council was persuaded that a letter ballot of the membership would give it even greater weight as a standard adopted by the American Society of Mechanical Engineers. Robert Thurston argued against that decision in a subsequent letter to the Council. He felt the Society should not take the responsibility for a standard, in the first place, and that as a matter of general procedure it was also dangerous to have committees speaking for the whole organization. Instead, the report

35 Ibid VI (1885) 256

should be accepted, with all its debate, and the committee discharged 'in the usual way.' Then, in a postscript, Thurston added what had been Kent's stand against recommending the Pratt and Whitney gauge: 'If the report be right, it will be accepted by the profession, whether endorsed by the Society or not; if wrong, the Society cannot help it, and will only be injured by its formal endorsement.'[36] That letter convinced the Council to rescind its earlier action and there the matter stood when it came before the members at the Atlantic City meeting of the Society in 1885.

Frederick Hutton, who had continually to remind his colleagues as the years went by that the Society's policy was to not adopt standards, wrote up the debate in a style that makes it seem ASME was arriving at a major decision, and historians have treated it as such ever since.

The discussion began on the question of submitting the code to the membership for its approval, but the issue was whether or not the Society would recommend compliance with a standard put forward by one of its own committees. E.P. Stratton spoke for a good many members when he claimed that some organization should adopt the standard and that none was 'better qualified' than ASME. Charles Emery, who had judged boiler tests at exhibitions of the American Institute in New York and at the Centennial Exhibition in 1876, argued that his courtroom experience as an expert witness led him to believe that formal adoption would do a great deal to resolve the legal disputes surrounding boiler tests, and he urged the Society to make a special exception in the case of this standard. John Sweet, old and much respected, put that motion before the meeting.

As the discussion went on into the morning, it became increasingly clear that formal adoption troubled a good many. That was the question he could not decide, George Babcock said, even though he 'would certainly like to see it become the standard of the country.'[37] The debate also made John Sweet wonder if he should not withdraw his resolution, which he did when discussion resumed in the afternoon. Still looking for some sort of compromise, however, he presented as an alternative that those present at the meeting 'recommend' the standard to members of the Society for their use, particularly in litigation. Kent, however, continued to press the argument that the organization should not 'put itself on record as adopting anything' and he was finally joined by Henry R. Towne, who reminded the members that opinions changed, 'and sooner or later, it will be sought to give the weight of the Society, as a body, to the enforcement of certain theories and certain

36   Ibid 879
37   Ibid 885

rules, which some of us may be sorry to see done.'[38] In the end, the whole subject of adoption was laid on the table.

It might seem as if the matter ended there, to the discredit of the Society's professional ambitions. Few of its activities, after all, better represented that urge for the systematic application of knowledge to the country's industrial development than the translation of research and experience into codified practice – to become, in turn, the basis of still more knowledge. And yet, when the moment came, it looks as if the Society failed to grasp the opportunity, out of some over-nice concern for the business interests of its members or from an undeveloped sense of social conscience. But it was the Society's ideologues who led the fight against adoption, not those representatives of 'shop culture' who wished for an untrammelled hand in the market-place. Furthermore, the history of standards in America is not only extremely complex, but almost totally unwritten and easy judgments are unwarranted.

Throughout the nineteenth century, Congress proved very reluctant to legislate any kind of standards, in the face of grave threats to the public welfare, and, even after the creation of the National Bureau of Standards in 1901, the formulation of uniform practices in engineering was still largely left to the private sector. In that arena of voluntarism, both the framing of standards and the way they became accepted were complicated issues. How ASME approached those problems depended on the nature of the standard and also changed over time. But from the outset, there was never any hesitation in anyone's mind about standards serving private interests – and there was no necessary contradiction perceived between private interest and public welfare. Self-interest was far from an opprobrious concept. Herbert Spencer, in Thurston's hands, had given it scientific sanctity and made it seem, in fact, an excellent mechanism for the regulation of human conduct. As ASME came to the business of standards, therefore, it did so from a philosophical position, as well as from a sense of human nature, that made interest perfectly compatible with fair play. So, for example, when the Society decided that pipe and pipe threads should be standardized, William Kent suggested the committee should be composed of 'men representative of pipe manufacturers and of pipe users, with perhaps one representative of the sprinkling system and certainly one of the manufacturers of taps and dies.'[39] That sort of approach to balance, which came to typify the make-up of ASME standards committees, neatly juxtaposed makers and users in a presumably offsetting

38 Ibid 888
39 Ibid VIII (1887) 20

fashion, while insuring that the committee also included the technical experience of other interested parties. From that point of view, it appeared reasonable to expect that in their own interest users would prevent the representative producer from measures that might give him an unfair market-place advantage over his own competitors.

The private sector context of standards formulation also insured co-operation with other societies and associations that had some interest in the matter. The Committee on Uniform Standards in Pipe and Pipe Threads conferred with similar committees from the Pipe Manufacturers Association, the Manufacturers Association of Brass and Iron, Steam, Gas, and Water Works of the United States, and with the Cast Iron Fittings Association, to make certain that the ASME standard would harmonize with the interests of those other organizations. Besides compliance and the rationalization of the domestic market-place, that sort of co-ordination also had as a major objective the expansion of United States trade, particularly in Canada and Latin America.

The same basic objectives characterized the move to devise standard dimensions for the flange diameters of pumps, valves, and allied apparatus that the Society started in 1887. Percy Sanguinetti, chief draftsman at the Franklin Sugar Refinery in Philadelphia, called for Society action that would 'result in the adoption of a uniform standard by the manufacturers.' His position was supported by W.O. Webber, a Boston consulting engineer, who claimed that if there were a standard manufacturers could send their products 'from Maine to California, Montreal to Chili, and be sure that our pumps, strainers, foot and flap valves would all fit and come out right.' There were disturbing elements of localism that stood in the way of continental markets; some manufacturers still looked to control the repair and replacement market through specialized fittings. But the members felt confident that the obvious need for reform and concerted pressure from the Society would bring the industry to a set of uniform practices. In a choice of words which suggested that the organization's policy not to adopt standards was far from settled, Webber moved that the committee's proposals for flange diameters 'be adopted and recommended as the ASME standards,' an expression the committee repeated in its own report a year later.[40]

The discussion surrounding subsequent standards reveals, in fact, that there was still a strong sentiment within the Society for ASME – sanctioned standards. But in the case of flange diameters, the committee's activities reflect another important dimension of the development of standards in the

40 Ibid ix (1888) 128

private sector. Since there was no mechanism for enforcement, compliance depended upon the co-operation of industry. That meant committees had to learn enough about existing practices to calculate a standard which was not so different as to retard its acceptance. As Sanguinetti put it in a circular he addressed to 48 manufacturers of pumps and valves, 'we do not think the changes we might ask you to make will be very expensive or troublesome.'[41] In a subsequent effort to secure uniform observance, a revised and expanded version was sent to more than 300 manufacturers and the committee joined forces with a standards group of the Master Steam and Hot Water Fitters' Association that was also interested in the subject.

By the development of standards that sufficiently satisfied the technical requirements of the case, without imposing large change-over costs, and through a growing set of co-operative relations with other technical societies and trade associations, the practical result of the Society's efforts at uniformity was that in fact they came to be known as 'ASME Standards.' The question of policy continued to come up at meetings, whenever the matter of standards was raised. It still seemed unfair to John Sweet to ask men to spend a good deal of time on a report and then have the Society refuse to adopt it. William Kent, however, could always be counted on to remind the members that they had long since decided against taking a stand on anything. But as the range and number of standards committees grew, so did the Society's sense of its power in this new and useful field of work. Thus, at the spring meeting in Nashville in 1888, in spite of the old policy, it was voted to appoint a committee to present a series of rules to be used in tests of pumping engines 'which could be followed by members of the Society in their practice, and could be recommended as standards in such tests by outsiders.'[42]

For all practical purposes, the question of Society policy was finally settled in 1895 and appropriately enough, the resolution came out of a reconsideration of uniform boiler testing methods – the subject of ASME's first standard. Time had made apparent some limitations in the original report and after a paper by F.W. Dean on 'The Efficiency of Boilers' a committee was appointed to consider a revised standard. But even though this was precisely the subject that led the Society to its policy in the first place (and exemplified Henry R. Towne's point that such a need for revision was the best argument against endorsement of a standard), the discussion of Dean's paper was full of references to 'The American Society of Mechanical Engi-

41 Ibid XI (1890) 590
42 Ibid IX (1888) 355

neers' Standard Code,' or the 'Code of Rules for Boiler Tests,' or the Society's 'Standard Code of Reporting Boiler Trials,' or most simply, the 'Society's Standard Code.' As Gus Henning put it, 'Everyone says that is the Mechanical Engineers' standard.'[43]

There was a natural attraction to standards that dealt with steam-boilers, pumping engines, flanges, pipe threads, and locomotives. Those things had directly to do with the professional practice of ASME's members and with the systematic operation of industries they owned or worked for. In most cases the standard aimed to bring about uniform practice in the United States, although sometimes it was framed with an eye to the larger markets of North and South America. The standards themselves were formulated less against an ideal than a set of practical realities that emphasized acceptance as the crucial issue.

During the time the Society was discovering its capability to shape American industrial processes, it also became involved in a program of *international* standards that stressed technical rather than business considerations, and appealed to the image of objective expertise instead of the interest of concerned parties.

The most vital of the early international standards activities concerned the search for uniform methods of conducting tests of the strength of materials. It grew out of the joint-society effort to secure government funding to continue the work of the Iron and Steel Board and was proposed by Professor Thomas Egleston, who had led that ill-fated campaign. Since there had recently been a European conference on the subject, however, it was evident this project should be international in its approach. William Kent, ever faithful to a limited mandate for standards committees, opposed the idea: 'the Germans and others on the continent of Europe, would wish us to express all our results of tests of materials in metric dimensions.'[44] But without the threat of compulsion or change-over costs, the metric system held no terrors, and at its Atlantic City meeting in May 1885, the Society easily appointed a committee of Henry R. Towne, Gus Henning, Thurston, Charles H. Morgan, and Egleston, who was named its chairman.

The work began slowly. As a first step the committee had collected samples of materials to be tested and had sent them to a number of institutions and business firms that had responded to a circular appeal for co-operation. That synthetic approach worked with items such as pipe threads, but

43 Ibid XVI (1895) 643
44 Ibid VI (1885) 17

Egleston's committee soon discovered that the test results they got back were completely unusable. Not only was there irreconcilable variation in the test procedures, but the reporting forms themselves differed to such a degree that the information they contained was also useless. Besides calling for a different approach to the committee's task, the results revealed, as Henry R. Towne said, that from the viewpoint of comparative utility, 'the immense amount of time and labor and expense which is being devoted to the investigation of material is largely wasted.'[45]

Much more fruitful to the committee's progress was the fact that Gus Henning's work took him abroad, where he was able to connect ASME's efforts with those of the Continent, particularly in Germany. He translated the proceedings of conferences in Munich and Dresden addressed to the problem of standardizing materials testing, and in the committee's first report he surveyed current work in Britain, France, Belgium, and Switzerland. From that experience and with a standardized form it had worked out to make possible the analysis of American data, the committee was able to frame a set of recommendations for those kinds of tests that could be conducted under highly rigorous conditions, the prelude to standard methods for routine shop tests. The World Engineering Congress, held during Chicago's Columbian Exposition in 1893, emphasized the value of international co-operation and gave extra cachet, just at a time when the Society was eager to entertain its foreign visitors, to this most intellectual and professional sort of standards activity. Henning, upon whose shoulders most of the Society's work came to rest, became its official delegate to the international conferences on materials testing, with his traveling expenses paid by ASME.

The quest for international uniformity involved matters of practical importance. Those standards concerned fundamental units of measurement and the basic construction materials of mechanical engineering. Furthermore, there were valuable economic implications in the ability to compare the quality of iron and steel from any part of the world. Goods like sheet metal and wire that were widely traded also called for standard gauges to describe dimensions. In other words, economic rationality as well as engineering efficiency demanded a common language of specification and a common understanding of the qualities as well as the quantities of the materials specified. But these kinds of standards activities also gave play to the intellectually appealing aspects of the problem. It became possible, in that expanded context, to search for a standard that had ideal qualities rather than one which must be made to fit existing market-place circumstances, possible to get

45   Ibid VIII (1887) 347

beyond the emotional fervor of metric system debate. In this fashion, international standards efforts helped legitimize ASME's entire standards program as a professional activity that systematized American practice to its economic benefit, and added in important ways to the world's stock of technical knowledge.

The maturation of ASME standards activity not only gave it intellectual respectability, but also suggested a political role for that work within the organization. This possibility emerged from the desire to reformulate the boiler test code, which raised the notion of a permanent committee for codes and standards. The boiler test standard is interesting for other reasons, too. It raised the idea of a permanent committee on codes and standards work.

Typically, Robert Thurston made that point, suggesting the possibility of 'constant revision' as a way of maintaining the most exact methods in a form understandable 'by every engineer of ordinary information and attainment.'[46] Here again was the essence of the Holley/Thurston program for ASME and for American technological pre-eminence. The concept involved both the use of a rigorous method to acquire precise data and a technical institution for the conversion of engineering science into industrial practice. Men such as George Stetson might employ Darwinian metaphors to describe the ideal process for the evolution of standards, or imagine an Adam Smith kind of context, in which market forces settled questions of usage. But ASME's ideologues envisioned a world regulated by expert technical knowledge, in which applied science determined the best way of doing things. In 1867 William Sellers had argued that the primary task of mechanical engineers was to devise methods and machines to increase the output of ordinary workmen. A generation later, Thurston claimed it was ASME's job to get up-to-date knowledge into the hands of ordinary engineers. In just the same way that he had calculated a complete system of technical education – from the elementary level where children learned to read by using words related to craft activities, to a post-graduate research and policy institute – Thurston fitted engineers into an industrial hierarchy, all the elements of which were aimed toward efficient production and technical advancement.

If ASME's standards activities were perfectly tuned to supply the demand of the mechanical industries for system and order, they also suited the creative needs of engineers trying to define a professional status for themselves. The formulation of a standard was a complex problem that called for consid-

46   Ibid 972

erable technical knowledge and practical understanding. William Kent, chairman of the first committee to present a set of rules for boiler testing, agreed with the need for its revision, but pointed out how difficult a job it would be, particularly since the 30 or 40 experts of a decade ago had grown to 300 or 400, and he predicted it would 'be a difficult matter to bring them together.'[47] But the growth of the manufacturing sector, and especially of the new electric utilities industry, that accounted for the increase in expertise, also made steam-power a subject of even greater interest with ASME. Besides the revision of boiler trial rules, the Society also took on the task of standardizing the methods of steam-engine tests – an effort that quickly assumed an international cast – and in conjunction with AIEE, began an investigation into the problem of standardizing the connections between steam-engines and dynamos. By the end of the century, to distinguish them from other committees, those groups working on standards were called 'the professional committees of the Society.'

This sort of activity lent itself to widespread participation, or at least to the possibility of it. Standards brought the Society into co-operation with a wide variety of organizations and in closer touch with its own members. Committee reports, for example, included standard forms to be used by anyone who wished to forward the results of his own experience in relation to the code. Thus, someone distant from the meetings, and hesitant about contributing written discussion to the *Transactions*, might still take a modest role in one of the Society's important technical programs. Perhaps it was a sense of that 'silent majority' which led Matthias Forney to argue that technical meetings of the organization might reasonably be held in any city, or that stimulated Gus Henning, whose standards work had given him a position of importance, to propose the creation of sectional groupings of the Society so that larger numbers of members might participate in its programs.

THE SOCIAL DIMENSIONS OF ENGINEERING PROFESSIONALISM

ASME's chief architects, Holley and Thurston, were concerned to design a new kind of professional institution to distinguish men whose mechanical knowledge led them into industrial pursuits. America no longer restricted its highest social esteem to those whose wealth flowed from land or commerce – the traditional sources of status. But for all the rhetoric that linked democracy and the mechanical arts, many of the Society's founders still resented the old and deep-rooted scorn of the liberal arts. Consequently, making a learned profession of mechanical engineering was a way simultaneously to

47 Ibid 1000

distinguish ASME's members from machinists, men who really worked with their hands, and to associate engineering skills with the intellectual expertise of the traditional professions. The Society came to life, therefore, with what might be called a social purpose and a technical purpose. One objective united the members in fraternal spirit, providing the matrix for a new social class, while the other gathered the power of expert knowledge and focused it on the solution of common technical problems. Both, in the minds of ASME's creators, were complementary aspects of professionalism, but as the Society evolved in its first two decades, they came to seem increasingly contradictory.

Except for a few dark moments at the beginning, it was always expected that the social side would prosper. In a description revealing for its style as well as for its content, Joseph Holloway once described the normal kind of ASME meeting, to contrast it with the 1885 Atlantic City gathering:

It has been our practice in the past, when we were able to devote a few hours to business and recreation, to seek some spot abounding in smoke chimneys and dusty thoroughfares, where, surrounded with rumbling wheels, roaring blast and hissing steam-pipes, we proceed to enjoy ourselves by climbing up and down cinder-heaps, tumbling over scrap-heaps and piles of pig-iron or rails, half blinded with the glare of roaring furnaces, and nearly melted by their heat. We came to the close of the day, begrimed with our surroundings, wilted as to our attire, but full of the conviction that we had been having a delightful picnic.[48]

Even in the unfamiliar character of 'beach loungers,' Holloway assured the mayor of that resort town, the mechanical engineers still looked forward to a good time. Thurston translated that self-image of hard workers and hard players into a description of the only kind of man who should be admitted to the Society – 'good mechanics by instinct, good men by original construction, good fellows by nature and habit and training.'[49]

Those ideas took on tangible dimensions in 1890 when ASME bought the New York Academy of Medicine's house on West 31st Street. It had a large auditorium for meetings, a room immediately below it for the late suppers that proved so popular, a library, an office for the Secretary, and even rooms on the upper floors for out-of-town visitors, just like a club. For years the Society had lived in rented space, into which it squeezed a business office that handled both membership and publications, a library that had been

48 Ibid VI (1885) 356
49 Ibid XIV (1893) 497

started in 1884, and a growing collection of portraits and historical memorabilia. In contrast, the new headquarters provided ample dimensions and a sense of established solidity. By 1890, the number of its members had lifted ASME to a position of equality with the other national societies and, in a way that clearly showed their enthusiasm, only 9 out of almost 1200 members were behind in their dues while the appeal for money to buy the Academy of Medicine's property was quickly over-subscribed.

There had been earlier attempts to acquire permanent quarters, but perhaps it was more than coincidental that the Society bought the house shortly after the excursion to Europe. Those who went abroad had not only seen the accommodation of European societies, but also returned with a new awareness of the social dimensions of engineering professionalism. The pleasure of travelling together on their chartered steamship and their generous reception abroad excited the feeling that they were part of a great international brotherhood, and that sentiment continued to echo in the Society's affairs after their return. It was the motivation for Henry R. Towne's proposal for an academy or institute of engineering that could represent all American engineers in reciprocating the hospitality the Europeans had shown. The same idea stimulated the World Engineering Congress in 1893, the movement for international standards co-operation, and, in 1900, the thought of a return visit to Europe.

Actually, however, most American engineers found their social interests best served by local engineering societies – such as those in St Louis, Chicago, Cleveland, or Pittsburgh – rather than by those national organizations located in New York. Indeed, the vitality of local groups during this period made it seem that in some form of coalition they might prove the most representative and useful professional institutions. William Kent saw in Towne's notion of an umbrella organization the means of combating this threat, by absorbing the municipal societies as 'chapters' in the larger structure, which would be controlled by the national associations. At the same time, ASME's leaders imagined that distant members might somehow find in the Society's new headquarters a more binding attachment to the organization itself, and that was another reason to solve the problem of housing.

Certainly the winter annual meetings, held regularly in New York after the house was acquired, became ASME's premier social event. The 1891 meeting, for example, opened with a *conversazione*, social reunion, and refreshments, featured a theatre party given the following evening by the local committee, and concluded with a full-dress reception and ball, a collation at 10:30, and dancing 'until an early hour.' The 'cosy' feeling of the Society's house, as Hutton often described it, led almost immediately to its use by local members for 'informal reunions.' These gatherings were inspired by such occasions as the unveiling of a portrait of Alexander Holley, or featured evening lectures on such topics as 'Robert Fulton,' 'Egypt, Old and New,' and 'Electricity previous to Galvani.' The Council was careful to point out that the expense of these informal meetings was borne entirely by those attending and that they were 'in no sense' official, but it nonetheless argued that they served important professional ends. The library also enjoyed a great success in its new quarters and by popular demand was kept open every weekday until 10:30 in the evening. Besides Holley's portrait and those of other important leaders, the Society acquired through donation Robert Fulton's original drawing of his steamboat, his dining table, and such things as John C. Hoadley's collection of test gauges and indicators, which gave evidence that ASME enjoyed cultural artefacts appropriate to its new setting.

Once the immediate details of moving in were completed, it seemed appropriate, too, to arrange the Society's financial affairs on a more regular and solid footing. Publication costs had doubled, along with the size of the annual volume of *Transactions*, and while most of the expense of meetings was met by local committees in the cities where they were held, the administrative costs of all other activities had increased. What the Council had done to meet these additional charges was to spend the income from life memberships and initiation fees, funds which admittedly should have been set aside for investment. Members had also been called upon to support some of the Society's activities through voluntary contributions, an approach that seemed unfair and added to the problem of keeping the accounts straight. The Council proposed to remedy this patchwork system by an increase in dues and initiation fees, which it submitted as an amendment to the rules at the 1891 winter annual meeting. The spirit of the amendment was perfectly in keeping with the feelings of established status which the acquisition of the new house had stimulated. Extra income from the increase would not only give a settled quality to the handling of finances, but allow for a surplus that could be used to support research, publish indexes of technical literature, or perhaps cover a portion of meeting expenses. And all of that could be achieved by a dues schedule still well within the means of 'professional men of the grade of outward success,' as Hutton described the membership.

The Society's leaders had forwarded the amendment as the fruit of their own matured judgment. 'The Council may be supposed, from its position as managing the business of the Society,' a circular letter to the membership began, 'to be better informed as to present wants and future possibilities than the membership at large.'[50] Nevertheless, it circularized the members for a straw vote on the proposed amendment, although with a set of choices wonderful in its phraseology. The increase in dues was described as a 'movement to enhance the value' of one's membership, while an opinion opposed to the amendment was stigmatized as 'a return to the earlier and lower standards of publication and management.'[51] But even such patent manipulation of language offended no one; by a ten-to-one majority ASME's members voted in favor of the increase. Long accustomed to an oligarchic form of government, those who attended that annual meeting in 1891 voted down another amendment which would have required the nominating committee to propose two names for each position, instead of the official slate it had always presented. C.J.H. Woodbury, one of the early vice presidents, spoke for everyone when he implied the amendment threatened 'the universal harmony which has always prevailed in the Society.' Furthermore, he claimed, 'the officers have always been closely in touch with the membership as a whole, and I want to ask you, in the name of common sense, considering all the results which have been achieved, is anyone going to be wise enough to better that matter?'[52]

Given the results, Woodbury's argument seemed a pretty good one. The dues increase swelled the treasury and soon enabled the Society to begin buying back the obligations it had incurred to finance the house purchase. By the end of the decade, in fact, the second mortgage had been completely retired. Annual meetings grew ever more popular and fell into a regular pattern that Hutton reported in a fashion which also became increasingly routine. The 'informal reunions' of local members in the Society's headquarters, that Council had so long protested were unofficial, took on a regular quality, too. But that came about as a result of an attack on their excessively social nature.

Matthias Forney, editor of *Railway Age*, had arranged a series of meetings on technical topics during the winter of 1894, modelled after monthly meetings of the Master Car Builders' Association, which had also been held at the Society's house, and at the annual meeting that year proposed a series of meetings for the following year 'for scientific discussion, and not occasions for jollification and social enjoyment alone.'[53] Actually, the informal reunions had already created some discontent, although it appeared in the *Transactions* in a

50   Ibid XIII (1892) 32
51   Ibid 35
52   Ibid 40-1
53   Ibid XVI (1895) 28

muted form. The essence of it was that such gatherings made the Society seem local in character and gave its headquarters the air of a private club for New York members, who also benefitted from easy access to the Society's meetings, office, and library.

Forney put his proposal forward in a way directly to engage that question of differential benefits. The organization was never 'intended to be a social club or trade union,' he claimed. Its central purpose was rather to allow technical men to exchange 'their ripest experience' and their 'most profound knowledge.' From that assumption Forney's point was that if the monthly meetings were technical sessions, they could not vitiate ASME's claims to national representation, and he suggested that such meetings might just as logically be held in cities other than New York. Forney's proposal raised serious issues. The Society was heavily biased toward New York City and aside from the *Transactions* and summer meetings it had no particular mechanisms to encourage the loyalty of more distant members. Furthermore, the dispute made it appear as if the Society's social and technical functions were not as complementary as the founders had imagined.

The debate on Forney's ideas did not, however, engage these problems. Joseph Holloway, whose business affairs had taken him from Cleveland to New York, sympathized with the plight of 'non-resident' members, and suggested that the monthly meetings should be organized under auspices other than the Society's. But Worcester R. Warner, another Clevelander, a partner with Ambrose Swasey in Warner and Swasey Machine Company, and soon to be elected president of ASME, set the tone for the discussion with the remark: 'I live out among the buffaloes and Indians of Ohio, and I know that several of our members out there, in arranging to come to this little village, look forward and plan to have their visits timed to occur when these meetings are held.' In the same mock-serious vein, Warner continued, 'We want to come down here once in a while, and as there is not much going on in New York, we want something to do of an evening.'[54] While the *Transactions* make it appear the issue was simply one of geography, Forney's point was that the social character of the monthly meetings made them unfair to non-resident members, and ASME's leaders did not like that implication. Within the confines of the Council meeting held immediately afterwards, Forney was rather pointedly left off the committee named to organize the kind of meetings he had proposed, and an upset Frederick Hutton urged that constitutional steps be taken to prevent those unplanned discussions of 'administrative business pertaining to the conduct of the Society or its policy,

54 Ibid 31

at times when consideration by those competent, is not possible or convenient.'[55]

The 'New York Problem' thus remained somewhat beneath the surface, with only occasional bubbles to suggest a reaction still underway. For the first time in fifteen years of reporting the Society's meetings, Hutton drew attention to the fact that the nominating committee was made up with geographic balance in mind. And when he became president, Worcester R. Warner took the occasion to remind those attending the 1897 annual meeting that the entertainment provided for out-of-town members had always been a personal courtesy of the local committee, and ought to be appreciated as such. There were other hints, too.

The 'Junior movement' was also concerned with geographical balance. An idea of Fred Halsey, who later gained fame in connection with Taylorism and scientific management, it had grown out of the monthly meetings as an attempt to encourage younger members to present papers before the Society. By 1899, the Junior Committee reported that an active program of monthly meetings was underway, and that they were especially occupied with ways to make the meetings 'of more than local interest,' and were working out a plan 'for reaching all distant members.'[56]

Halsey was one of a younger group of men rising to prominence in ASME and the Junior Committee consisted of younger members, too. The tone of their report was new within the Society and seemed to suggest criticism of the existing order. The movement, they claimed, involved much more than the simple business of bringing young men into the organization's work. It was a matter of engineering education, of raising the standards of the engineering profession, and indeed, of 'the future of the National Society of Mechanical Engineers.'[57] As Forney had earlier realized, the social element of professionalism thus appeared to intersect with regionalism, on one hand, and with an authoritarian power structure, on the other. But this modest challenge soon faded. There were Junior meetings for a year or two, although what tended to happen was that senior members came to dominate them. And there were other problems with the movement which the *Transactions* do not describe. Instead, there are allusions to the need for a 'differently constituted body,' and to 'inherent difficulties' that ultimately led the committee unanimously to resign in 1901.

55   ASME Council Minutes 7 Dec. 1894
56   ASME *Transactions* XX (1899) 454
57   Ibid 453

These hints of trouble were nothing more than that. The Council tightly controlled the information appearing in the Society's publications, particularly anything that was controversial. Besides, out of their business experience and their sense of professional behavior, engineers were normally discreet about such things. And in any event so much else seemed in good order.

THE REVOLUTION OF 1901

ASME's founders had imagined two great fields of activity for the organization. In one, the play of social forces would effectively mix business and pleasure; in the other, the drive for industrial efficiency would combine business and knowledge. These central objectives were seen as essential and interrelated elements of professionalism, somewhat in the way that a healthy life integrated work and play. Professionalism was thus to be an expression of intellectual and business interests, but it was also to serve as a means of making engineers more aware of their common political, economic, and social ambitions. After two decades of experience, however, that neat, symmetrical, and interlocking conception seemed to have broken down.

To the outside world ASME presented a picture of almost complete success. With a large membership, comfortable house, considerable income, and important men in its highest offices, it appeared the ideal instrument for the twentieth century. The Society's traditional strength in the machine-producing industries made it the spawning ground for new automobile and aviation technologies, while its long connections to steam-power linked it to the electric utilities industry, another emerging giant. Besides these sources of expertise, the Society had also discovered that management logically fell within its sphere of influence and the rational organization of work indicated another role of enormous future importance. On top of that, ASME embodied in institutional form two central tenets of the national faith – that technology was the natural artistic medium of Americans and that democracy and material abundance were joined in an inevitable historic role. But despite all those apparent advantages, the Society's treasury was, in fact, empty, its administration inept, and large numbers of its members alienated by an authoritarian Council. On the inside, although the *Transactions* scarcely suggest it, the discontent grew steadily and explosively until it finally erupted, at the annual meeting of 1901, in a bitter fight over the way the organization was governed.

The actual issue that triggered the upheaval was a proposed dues increase, notice of which had been given at the previous spring meeting, as called for by the rules. The language of the proposal was astonishingly like that used in 1891 to justify an increase then: in order to meet increased expenses, it had been necessary to spend income from initiation fees and life memberships; however, if a dues increase were approved, the Society could meet its obligations, fund research, and publish indexes to engineering literature. But again, the debate did not even touch those arguments. Instead irritation was concentrated at four sore points: Council's high-handedness, ASME's administration, the geographic imbalance in its program, and a rather vaguely identified resentment of its luxurious style.

In the first place, the Society's financial problems took the membership entirely by surprise, and they were as much angered by Council's lack of candor as they were by the news itself. 'Let our Council be open and above board with membership, and tell us what is going on,' one member urged, 'let them cease trying to play the "paternal" act over us, for we do not want it, neither will we have it.'[58] But the discussion revealed that most of the Council had been equally ignorant of the financial problem and that naturally focused attention on Frederick Hutton. To men accustomed to thinking in terms of the economies of scale, the most obvious question was, 'Why should it cost more *pro rata* to run a society with 2,000 members than one with 1,000?'[59] It was just as clear that the Council had no answer and that a special committee would have to unscramble the problem.

It seems strange, in retrospect, that a dues increase should have been proposed without a detailed explanation, particularly when it would have made ASME's dues higher than those of any other American engineering society. But the leaders were as much out of touch with the membership as they were with the finances. When the idea of an increase was put forward within the Council, its members had no way of knowing a strong resentment already existed among the rank and file over just such cavalier attitudes. And not only was there no mechanism to discover that feeling, but Council members also had not usually thought to ask the membership for its advice in any event. Instead, the Council planned to announce the proposed vote on the dues increase at the Milwaukee meeting in May 1901, and have the vote the following November in New York, where it probably expected to have a majority.

But the technical press fastened on the issue and succeeded in forcing the Council to accept proxy votes from members unable to attend. *American Machinist* and *Railway Age*, in particular, became the voices of a widely scat-

58 Ibid xxiii (1902) 39
59 Ibid 58

tered constituency that saw the dues increase as a move of New York City members to benefit from an ever more extravagant headquarters operation. As one of them put a common complaint: 'The clerical force now numbers four, I believe, while the secretary spends but little time at the society house. It seems to me that half this force should do the work. The expenses seem to increase faster than the membership.'[60] The difference in incomes between those very wealthy members of the Society and the ordinary engineer was also alluded to in the letters that poured in to the editors of technical journals, and it cropped up in the discussion at the annual meeting, too. 'I am not wealthy like a whole lot of men in our Society,' W.S. Rogers said, comparing himself to 'the others of you who can throw away $1,000 to my penny.'[61] Those kinds of sentiments were a clear indication of the dissatisfaction a great many members felt over the conduct of the Society's affairs, and when the votes were counted – Matthias Forney held over 100 proxies, all of which he cast in the negative – the dues increase was defeated by a three-to-one margin, with about half of the members voting.

That was far from the end of the matter, however. At the following spring meeting in Boston, criticism of the administration grew even sharper. Most of it came from Fred Halsey and Harrington Emerson, another advocate of scientific management. Neither minced words. The same training and experience that qualified them for admission to the Society, Halsey argued, entitled the members to know and participate in the organization's business. They should be told everything about its finances, including the cost of running the Secretary's office compared to what other societies spent on administration. And not only should there be an end to secrecy, Halsey claimed: 'There is quite a prevalent and perhaps growing conviction that this Society will never be able to do all that it should do; nor to develop the degree of usefulness it should develop, or might develop, until it has a Secretary who is not only efficient and business-like, but who can devote his entire time to the affairs of the Society; and will have no other interests which he may consider paramount.'[62]

Emerson also centered his attack on the inefficiency of the Society's administration. In the same way that Louis Brandeis later argued that railroads could solve their financial problems by cutting waste rather than increasing rates, Emerson claimed that ASME could easily pay its debts by curtailing expenses and streamlining its publication procedures. It was at

60 *American Machinist* (21 Nov. 1901) 1301
61 ASME *Transactions* XXIII (1902) 39
62 Ibid 417

least ironic, he pointed out, that a society claiming to have America's best organizing talent in its membership, should be 'one of the most inadequately managed' he'd ever seen![63]

Not all the criticism came from new disciples of scientific management, however. Some of the Society's old and loyal members took the opportunity to air some of their own concerns, especially about the *Transactions*. William Kent protested that many of the papers that had been published were no more than commercial advertisements, 'unfit to print,' and a discredit to the organization.[64] Gus Henning joined him on that subject, as did Colonel E.D. Meier. It was difficult to get enough papers of suitable quality, but those that were obviously inappropriate were not always rejected – another charge laid at Hutton's doorstep.

It was clear that ASME was headed for a period of substantial reorganization. And in the debris of the dues increase fight, it was also evident that there were discontents other than that over Hutton's ineffectiveness. Indeed, that may have been the least of them. He was rather more a symbol of a certain style that made the New York headquarters seem to distant members a place of self-indulgence. This reaction was not a piece of morality, pitting the virtues of rural America against the luxurious idleness of the metropolis – most mechanical engineers worked in urban centers. Instead, what stirred the 'out-of-town' members was the sense of an exclusive clique – by which they meant the closed methods of Council as well as elaborate social events – and the feeling that the Society was not living up to its promise as an association of professional men, particularly in its obligations to the country at large.

Henry R. Towne, who had played such a calming role at the annual meeting, when tempers were short and there seemed the real danger of a serious rupture, took on the same statesmanlike mantle in Boston. A past president, and long identified as a member of the inner circle, he had also inaugurated the movement to include 'economic' subjects in Society meetings that paved the way for Frederick Winslow Taylor's first paper on scientific management. Towne thus enjoyed status with the establishment and credibility with the efficiency-oriented group of reformers in the organization. From that position he was able to propose an interpretation of the fight which not only contained elements of truth but put the best possible face on events. It was, he claimed, 'the most profitable' meeting he'd attended in a long time. The Society had fallen into comfortable ways: 'The members have been leaning too heavily on the Council, the Council has been leaning too heavily on the

63 Ibid 424
64 Ibid 427

Secretary,' and committees had been left too much on their own. But it was the members themselves that determined the organization's health, Towne observed, the members that gave an organization its vitality and impulse. What ASME needed was a '*reawakening*': 'We want to get back again to the spirit we had in the earlier years.'[65]

By 1901, most of the essential elements of ASME's subsequent history were in place. The role of staff in a volunteer organization was, for example, an obvious issue growing out of the Society's first two decades of existence. So was the relation of the New York headquarters to a membership widely scattered over the North American continent. Mixed up with that problem was the matter of the Society's social and political style. An oligarchic form of government came easily to men whose business relationships were predominantly hierarchical and to a profession that esteemed business success so highly. But as the organization's membership grew, as the *Transactions* expanded, and as technical activities increased, the Society could not be effectively managed by a small group of insiders. Besides, the old approach did not sit well with everyone – professionalism also carried with it the concept of people made equal in standing by their expertise.

The Society founders had envisioned its social and technical programs as interrelated and mutually reinforcing. But twenty years of experience cast that hope in doubt and made less clear than it had seemed at the beginning just of what each of those elements consisted. It looked as if ASME's technical activities might even prove more effective in knitting together a community than its cotillions and *conversazioni*, although in the arguments of the reformers there was the idea that social affairs might be redefined to mean the general interests of members and the relation between engineers and the public. Whether the Society could learn to be as candid with the public as its members asked the Council to be with them was also a problem for the future. But even in 1901 it was clear that how ASME handled information about itself and about mechanical engineering was the most crucial issue in its history.

65 Ibid 430–1

ROBERT HENRY THVRSTON
DIRECTOR
1885   1903

·FREDERICK·REMSEN·HUTTON·
·SECRETARY·OF·THE·SOCIETY·1883–1906·
·PRESIDENT·1907·

Office and parlor of the building, on West 31st Street in New York,
which ASME used as its headquarters from 1890 to 1906

LEFT, TOP Robert Henry Thurston (1839–1903)
An exact replica of this 1908 bronze memorial to Thurston at
Cornell University was made for ASME's New York headquarters where,
with appropriate ceremony, it was dedicated in 1910.

LEFT, BOTTOM Frederick Remsen Hutton (1853–1918)
A bronze memorial tablet recognizing Professor Hutton's long service as
ASME Secretary also once decorated the Society's headquarters.

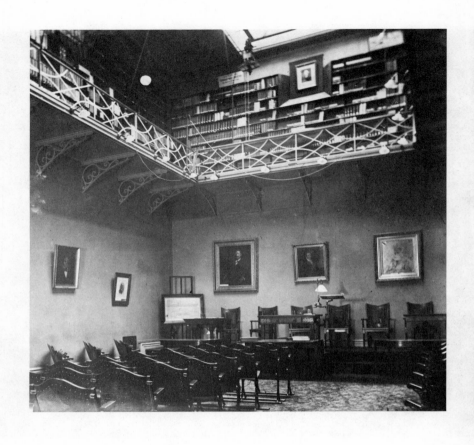

Auditorium of the West 31st Street building

This picture, taken at a Council meeting at the 1903 annual meeting,
is remarkable for the number of presidents it shows.
James Mapes Dodge, the current president, is seated at the head of the table.
To his left are Frederick R. Hutton, John E. Sweet, Henry R. Towne, Oberlin Smith,
Charles H. Loring, Charles E. Billings, John Fritz, and Worcester R. Warner.
Standing behind them, from left to right, are Erasmus Darwin Leavitt,
Charles W. Hunt, Admiral George W. Melville, Samuel T. Wellman, and
ASME's long-time treasurer, William H. Wiley.

3

ASME's Social Economy

What ASME's founders distinguished as its social purpose and as its technical purpose became two great currents flowing through the Society's history. One expressed the application of rigorous training and specialized knowledge to the solution of technical problems. The other, a less natural analytical category, reflected the engineer's desire for social status and for political and economic power. Over the years, these elements have been mixed in various and often contradictory combinations, and have assumed forms that differed substantially from one time to another. But perhaps in the way that the search for identity describes a kind of continual force in individual psychology, so the interplay of its social and technical characteristics has from the beginning defined the essence of ASME's nature.

Social purpose most obviously suggests entertainment, the recreation at winter and summer meetings that was to relieve mind and spirit after the technical sessions were over. But Thurston had a much broader range of implications in mind. When he spoke of 'the field of social economy' that lay before the Society, just as important to him as the technical work to be done, he meant the concerted power and influence of the members applied to politics and to public opinion in order to advance their business interests and professional standing. The social element thus encompassed all the Society's non-technical relations with the outside world. But even more important, as the first twenty years of experience proved so conclusively, the social element also included ASME's governance and raised most crucially the question whether the Society actually represented the hopes and ambitions of mechanical engineers throughout the country.

All of America's major engineering societies had to struggle to make good their claim of national representation. But for the ASME it was especially difficult. The Society, even more than the others, was oriented to New York

City. More of its members were based in the city and its immediate vicinity and more of the industrial firms mechanical engineers worked for had their headquarters there, too, drawing additional members to the metropolis on frequent occasions. Thus, unlike AIME, which in its early days held its meetings in cities near mining districts, ASME slipped naturally into a mode of operation centered in New York. It was always more convenient to make committee assignments and frame nominations with easy access to the city in mind, pleasant to have monthly gatherings there, and always easier to conduct business informally at the Engineers' Club, particularly for an organization with a small membership, most of whom knew each other. In those respects, ASME was more of a local society than a national institution before the 1901 revolution.

One of the Council's arguments in that ill-fated campaign for a dues increase had been that the Society could not continue to expect such a rich harvest of initiation fees. But in a piece of institutional ignorance allied to their failure to anticipate the turmoil, ASME's leaders had not recognized a set of important occupational and geographic changes in mechanical engineering already underway. In a manner that paralleled the growth of industry and of urban populations, the total number of engineers in America shot up dramatically, swelling by almost 2,000 per cent in the years from 1880 to 1920. As a result, the Society, which had been dominated by men who owned and managed their own firms, became increasingly populated by men who were employees of such firms, and their number continued to grow at a rate that also increased with time. Another of the changes was that more and more of these new men worked outside New York City. And as large corporate structures ruled an even greater proportion of the mechanical industries, engineers in those bureaucracies began to seek other ways to express their ambitions for professional status. One method was to discover in local societies some of the satisfactions New York members enjoyed at ASME's headquarters. Another was to make the Society more truly representative of the geographic spread of its membership.

ASME's leading figures had always been worried that a New York focus might make the organization seem local. It was that sort of concern with the appearance of things which led the Council in 1891 to propose a meeting in San Francisco, 'to put beyond cavil the national character of the organization,' even though that location prevented 95 per cent of the membership from attending.[1] But the Society had acquired its house by then, exacerbating the

1 Council Minutes 16 Jan. 1891

geographic problem, and the Council was eager for proofs that it spoke for all of America's mechanical engineers. In fact, ASME was actually in the process of becoming a national society with diverse technical interests and a widely scattered membership, and the Council was soon forced to deal more directly with these centrifugal tendencies. On one hand, a steadily increasing membership provided the basis for a larger and financially more powerful association; on the other hand, however, those same tendencies threatened the oligarchic form of control the Council exercised over rank and file members. The problem of striking a geographic balance in its affairs became, therefore, one of the most important themes within the social sphere of the Society, touching practically all the other significant issues in its history.

The model for the sort of unit that members in other cities might form was provided, ironically, by New Yorkers. From the outset there were metropolitan members opposed to a combination of social and professional functions, but ASME's new headquarters was drawn irresistibly into just such usage. A house-warming party in 1890, at which time Alexander Holley's portrait was unveiled, presented one such occasion and in the natural course of things it was followed by other 'social reunions,' although the Council made it clear they were to be paid for by those attending. These 'ASME evenings,' as they were also called, came to be held on a monthly basis and, while they may have begun as purely social events, it was not long before some of them took on the quality of technical meetings although none of their papers ever appeared in the *Transactions*.

It is difficult from the available historical sources to discover much about either the character or evolution of these meetings, but they proved controversial in whatever form. The Council continually assured members that the cost of such unofficial affairs was entirely 'independent of the Society's funds,' yet it was clear that they were organized under its auspices and in 1892 an editorial in *American Machinist* called for separate and distinct monthly meetings of mechanical engineers, the proceedings of which might be available to out-of-town members. This attack, followed by one from *Railway Gazette*, came from younger New York engineers who objected to the fact that the local meetings were dominated by an ASME headquarters group more interested in social gatherings than in technical exchange.[2]

In a move that might have created the basis for a metropolitan section of the Society, Matthias Forney successfully outflanked the Council at the 1894 annual meeting with a resolution that established a committee to plan regular meetings for the discussion of 'subjects pertinent to mechanical engineer-

2 *Railroad Gazette* XXVI (1894) 95

ing.'[3] But no one was quite ready yet for an extra layer of organization and the plan was ignored. Instead, some attention was given to the idea of a differential dues structure, like that of Britain's Institution of Civil Engineers. Those members of the Society who lived within a given radius of the New York headquarters would pay higher dues than those further afield, and it was implied by some that out-of-town members could use the difference in fees to join an engineering society or club in their own locality. However, that idea was not very attractive either, even though New York members clearly received greater benefits from their dues than those unable to use the library and meeting rooms. The Society had to have a headquarters somewhere and New York was the most sensible location for it, but it always seemed that a large proportion of members paid for facilities which a much smaller percentage enjoyed, and the Society continued to struggle with the problem.

Forney's idea of monthly technical meetings was subsequently taken up by Fred Halsey who converted it to the purposes of his own reform movement on behalf of the Society's junior members. His plan did not prove any more long-lived than Forney's, but both were important steps in the gradual process toward official recognition of local meetings and their integration into the Society's administration. Gus Henning brought another perspective to the problem. His experience abroad had made him familiar with the practices of European engineering societies, and in the *Verein deutscher Ingenieure*, with its local sections scattered throughout Germany, he saw a solution to ASME's financial problems as well as the geographic one. Henning presented a thoughtful plan for local sections at the 1901 annual meeting, but in the uproar over the proposed dues increase, his ideas did not receive much attention. Henry Towne did not fare better at that meeting with the suggestion that the Society take it upon itself to organize monthly gatherings for local members at its West 31st Street headquarters. Two years later, the Council considered Towne's idea but its members were still worried that people at such meetings 'might claim the right to transact Society business.'[4]

ASME's Council had never been comfortable with the thought of sharing its power and the 1901 revolution against its authority – as well as its New York bias – made it all the more defensive just as the movement for local sections was gaining strength. Those opposing forces within the Society produced contradictory effects. In 1905, for example, the Council finally passed a set of

3   Council Minutes 7 Dec. 1894
4   Ibid 9 Dec. 1903

rules providing for local branches, but then squashed the efforts of a group of Milwaukee members to establish an ASME section according to the guidelines, because they appeared to have initiated the action themselves rather than awaiting Council's permission. That strict interpretation, *American Machinist* reported, 'caused the Milwaukee branch to dissolve as such and become a local engineering society.'[5] And the rules themselves were hardly calculated to encourage the creation of sections; local groups were forced to assume all the responsibility for their financial affairs but to turn over control of all their technical papers to the Society, which also had the power to suspend or disband a section on sixty days' notice.

In later years, the Society would list the Milwaukee section as its first, founded in 1904, but it was in fact another six years before the legitimacy of geographic divisions was even secured. The problem was common to the entire field of engineering professionalism. Just as the American Society of Civil Engineers had opposed the formation of other societies on the grounds that it represented all non-military engineers, later organizations also fought the subdivision of their memberships and spheres of influence, whether through the splitting off of technical interest groups or through the creation of regional associations. And in 1905, when engineering's institutional arrangements were still far from resolved, local sections seemed as much a threat as an asset. An *American Machinist* editorial in 1905 commented that throughout the profession the matter of local sections was 'receiving a great deal of attention,' and described the movement as perhaps the only way members distant from New York could effectively participate in a society. The editor did not argue any need for structural relationships, but rather imagined the evolution of sections as 'a question of the survival of the fittest.'[6] Of course, it was exactly the administrative connections that needed to be worked out, and they emerged not from some budgetary struggle for existence, but from ASME's gradual accommodation between the pressure of distant members for geographical representation and Council's urge to concentrate its own power.

The meetings committee, which in the past had been responsible for the annual and semi-annual meetings, became the delivery vehicle for local sections. During 1909, without much official notice, local activity had begun in St Louis and Boston – the location of the two oldest local engineering societies in the country – under the control of the meetings committee, which had earlier taken charge of the New York monthly meetings. That arrange-

5 *American Machinist* XXVIII 2 (1905) 813
6 Ibid 68–9

ment allowed ASME to relieve some of the pressure for local sections while also claiming that 'all meetings, wheresoever held, are meetings of the Society; they are not local meetings of branches or sections.'[7] But to whatever degree the Society wanted to preserve its control over meetings in other places, only a little experience was enough to make it appreciate the difficulties of such long-distance arrangements, and in 1910 the Council's Executive Committee directed the meetings committee to review the matter. The report of a special committee on policies the following year recommended that members in any city be allowed to organize local meetings, without reference to the meetings committee, and that the Society assume some of the costs of those meetings. At the same time, publications procedures were also reformed 'to increase and extend the benefits of the Society to members at a distance.'[8]

Those actions in 1911 practically ended the Council's opposition to the creation of ASME sections. Only a month later it gave local committees authority over their own meetings and at the Society's annual convention the following November a formula for financing local activity was devised that also allowed section membership by those who were not ASME members. That had been a sore point in the Milwaukee controversy and, although it was momentarily resolved in favor of local groups, it remained a troublesome issue until the late 1930s when the Council finally returned to the position that section membership required Society membership.

In this rather deliberate fashion, ASME arrived at the position by 1912 that local sections were a positive benefit. As part of that process, the Council arranged a conference in the spring of 1912 to consider the form of local sections and their relation both to other local engineering organizations and ASME. The conference reported a set of proposals – originally outlined by Ernest Ohle of the St Louis section – describing the relation of sections to the parent body, that proved to have enduring significance. For all the behind-the-scenes tension the subject had created, Ohle's guidelines were remarkably simple and straightforward. The Society committed itself to the encouragement of sections and to the return of a portion of membership dues for their use. The sections themselves were established as geographical units of the Society with the power to elect their own officers and organize their own meetings. These mutually beneficial arrangements gave sections the opportunity to ally themselves with other local engineering groups and

7   Council Minutes 31 May 1910
8   Ibid 10 March 1911

connected distant members to the Society in a workable fashion. Ohle's suggestions also called into existence an administrative mechanism, the Conference on Local Sections, to carry out the program.

As the Council originally described it, the conference was to be made up of two delegates from each city interested in organizing a local section, a plan that suggested a search for grass roots judgments. But in actuality its membership consisted of the chairman, Professor Ira Hollis of Harvard, representing Boston; Professor William F. Durand of Stanford, representing San Francisco; Colonel E.D. Meier, of St Louis; Robert Yarnall, of Philadelphia; and two New York representatives, J.W. Nelson and Fred Colvin. That group of Society notables, four of whom were elected to the presidency, signalled a trend of great future importance. Geographic divisions were the new source of political power in ASME, although that did not mean a diminution of New York's authority. Indeed, as it turned out, the local sections movement created a mechanism for ambitious men with a geographic base of support to rise to the Society's highest offices. For example, the 1912 conference of local sections led to regular conferences of section delegates at annual and semi-annual meetings after 1915. From the Society's point of view, the purpose of these conferences was to teach delegates ASME's administrative procedures and to acquaint them with members of Council and with 'the ideals which guide this important body of the Society.'[9] For those interested in politics, however, the conferences were an ideal school for advancement.

In time, these conferences also took over the nominations process, and that served the interests of men from the Society's geographical divisions, too. Once tightly controlled by Council, the Society's procedures for electing its officers were now jarred by the same dissention over a New York – dominated, authoritarian form of government that had led to the 1901 dues revolt. But it was still some time before the Council was willing to do more than entertain opinions from the membership about official nominees. As Calvin Rice discovered when he became Secretary of ASME, its leaders generally took the position that 'no nominating committee should be nominated until a few in the Society know pretty well whom they are to nominate.'[10] In 1916, however, President David S. Jacobus invited sections to suggest names for membership on the nominating committee and that idea proved so successful that two years later the whole nominating process was placed entirely in the hands of sections, grouped into regions 'along lines similar to

9   Ibid 6 June 1929
10   Calvin W. Rice to Frederick W. Taylor, 11 May 1907; Taylor Papers, Stevens Institute of Technology, Hoboken, NJ

the American Society of Civil Engineers.'[11] Thus, the sections produced the men who selected nominees and the result inevitably favored men with strong regional popularity.

Nothing better illustrates the case than the Committee on Local Sections, which became a veritable nursery of presidents. No other standing committee produced so many officers of the Society. Of the 24 men who served on the committee from its establishment in 1914 until 1929, for instance, seven rose to become president, six were electd vice president, two became managers, and one the chairman of the finance committee. And that enumeration only considers the highest office each attained, not the various offices each filled during the same period.

From the Society's records it would appear that by 1917 the local sections problem was largely resolved. The Council had approved a new set of by-laws governing sections and at the annual meeting that year Robert Yarnall, chairman of the Committee on Local Sections, emphasized the freedom of sections to co-operate with other local groups and to organize their own affairs as they saw best. The format of the session itself, during which each delegate spoke briefly about his section, was also designed to illustrate the potential variety in their activities. There had been more than 150 local meetings that year, Yarnall reported, scattered all over the country, but as a reflection of headquarters' concern, the President, Ira Hollis, had visited 'practically every Section of the Society.'[12] The picture Yarnall drew was one of healthy development, with sections encouraged to a sturdy independence by a Society whose officers exercised themselves in fostering care.

Louis C. Marburg, a member of the Committee on Local Sections, addressed the Conference of Local Section Delegates at that same annual meeting on the effects of the new by-laws, and he echoed the theme of mutual benefit. The rules not only permitted non–ASME members to join sections, a major demand from local groups, but also gave them control over the dues of those affiliates. In particular, Marburg stressed the need for co-operation among engineering societies. 'Engineers in districts remote from New York,' he said, 'require an organization nearer to themselves than the national society, nearer not only in distance but in interest.'[13]

Marburg's remarks, printed in the Society's *Journal* (later renamed *Mechanical Engineering*) attracted considerable attention. In a subsequent issue of the *Journal*, Charles Whiting Baker, editor of *Engineering News*, a

11 Council Minutes 18 Jan. 1918
12 ASME *Journal* (Jan. 1918) 80–5
13 Ibid (March 1918) 209–11

former ASME vice president and long a liberal voice in the profession's politics, reiterated the need for co-operation rather than competition among engineering societies and pointed out that all the national organizations were creating structures for better geographic representation. Irving E. Moultrop was another of those who wrote in to praise Marburg's talk. A Boston Edison Company engineer, active in the city's engineering society, and chairman for several years of ASME's Committee on the Increase of Membership, Moultrop knew the issues that touched local engineers. The local sections movement meant to him a more democratic society and made it clear 'that our Society is for the benefit of the engineer, no matter whether he is located in New York City or in some remote part of the country.'[14] But the question of geographic balance within the Society was not one that resolved itself into simple moral categories.

Frederick Winslow Taylor's election as president in 1906 gave hope to the reform wing of the Society on several counts. He was elected to office on the assumption that he would regenerate a Society whose administrative arteries had hardened and which had proved unresponsive to the economic and social interests of younger engineers located outside New York. Scientific management promised a new, more objective kind of engineering professionalism that appealed to many younger men, just as they anticipated that Taylor's methods would streamline ASME's internal operations. Central to the reform process, however, was a broadened base of participation. Constitutional revisions in 1904 had created new standing committees to achieve just that, and Taylor recognized very clearly the need for an infusion of spirit into the Society's affairs. But as he came to fill committee posts with the sort of men who would put time and effort into the job, he discovered a set of difficult realities.

Besides the fact that the most energetic men were already busy, those living at any distance from headquarters were simply unable to attend committee meetings with any regularity, and business was often forestalled for want of a quorum. That result, ironically, stemmed from an effort to staff committees with men from different regions. Taylor also felt pressure on him to appoint representatives of the various branches of mechanical engineering, and he learned that, like the offices of vice president and manager, committee assignments were sometimes seen as a means of rewarding the faithful. He therefore made it a condition of his own reform of the Society to pick those 'best suited to the work' and 'to appoint on our important com-

14 Ibid (June 1918) 467

mittees only members who are in comparatively easy reach of New York City.[15] If nothing else, then, efficiency meant that the Society would continue with a New York bias, to the discomfort of some members. But as previous experience had made clear, the geography issue was never separate from questions about the Society's administration, its political style, or about the relation of engineering professionalism to larger questions of public interest. Scientific management gave a particular focus to all those tangled themes.

The reform spirit in ASME thus achieved establishment status in 1906 with Taylor's election. Taylor was the ideal person to lead a movement for the regeneration of the Society. A Vice President since 1904, he enjoyed an outstanding technical reputation, a secure social position, wealth, and energy. Furthermore, it had become obvious that the Society's affairs were badly handled. Hutton, using methods essentially unchanged since 1883 when he had become Secretary, enjoyed the sense of controlling the Society's affairs and it was not easy for him to delegate authority. But committee and Council members were sometimes glad to have him do their work and that also prejudiced the effective administration of the institution. Furthermore, Hutton already had a full-time professorship at Columbia University and that further reduced his ability to manage an increasingly complex organization. Taylor took on the presidency partly because everyone looked to him to straighten out the Society's administration, and he must also have been attracted by the idea of applying his principles of scientific management to ASME itself. But, along with his younger disciples in the organization, he also saw the need for reform in order that engineers might play a more prominent role in American society. His tasks were clear; ASME needed to find ways more actively to engage members in its affairs, its business procedures needed a thorough reorganization, and it needed the infusion of new ideas and vitality in its administration.

Henry R. Towne was chairman of the nominating committee that selected Taylor and in his characteristically diplomatic way he described the Society's situation: 'It has been run for a long time on established lines.'[16] As incoming president, one of Taylor's first tasks was to appoint members to standing committees and he did so with a vigor that ratified Towne's choice and marked the other features of his tenure. He recognized, as he put it to a prospective appointee to the membership committee, that 'the future success of our society will depend more upon the efficiency of our committee

15   F.W. Taylor to E.I.H. Howell, 10 Dec. 1905; Taylor Papers
16   Henry R. Towne to F.W. Taylor, 20 Sept. 1905; Taylor Papers

members, perhaps, than upon any other one factor.'[17] Taylor's vision was to populate committees with hard-working and conscientious members, identify the work that best employed their skills, and define the jobs most suited to headquarters staff. And his approach was more than an effort at the rational division of labor – it aimed to reduce the Secretary's authority and increase the members' responsibilities.

Hutton saw the matter in that light, too, and in a counter-attack drafted his own plan for the operation of standing committees which differed 'materially' from one the Council had outlined. In effect, it was nothing more than his customary mode of proceeding, reworded to suit the times. Each committee would make a single annual requisition for the money it would need from the coming year's budget, leaving the administration of it to the Secretary and his staff. The problem, as outgoing president C.W. Hunt pointed out in a letter to Taylor, was that 'it takes away discretion and management from the Committees, where it should remain.'[18] But the incoming president was firm in his conviction that the job of implementation be taken off the shoulders of committee chairmen, without shifting authority and responsibility to the staff, and that was one of the reasons Hutton felt slighted by the changes Taylor brought in.

Applying the principles of scientific management to the Society that prided itself for having spawned the new science was probably the most famous of Taylor's reforms, and he was not long in getting that project underway, as well. He outlined his approach in a circular sent to members of the Council before their first meeting with him, suggesting the formation of a committee staffed by a special assistant in the Secretary's office, 'whereby detailed study might be given to possible simplification and standardization of the office routine.'[19] Behind that deceptively mild language, however, was a plan for the complete renovation of the Society's procedures. Taylor's 'code of standards' for ASME would spell out the duties of the president and of the standing committees and define their relation to headquarters staff. In the same way that he approached industrial firms using his methods, Taylor wanted to extend his system to every aspect of the Society's business, whether touching its publication processes or the conduct of its meetings.

The magnitude of the job argued for a person solely responsible for it, and to make that conclusion easier to swallow, Taylor offered to pay the special assistant's salary. In that artful fashion, he was able to have his own man

17 F.W. Taylor to E.I.H. Howell, 10 Dec. 1905; Taylor Papers
18 C.W. Hunt to F.W. Taylor, 17 Jan. 1905; Taylor Papers
19 F.W. Taylor to ASME Council 26 Dec. 1905; Taylor Papers

working directly in the Secretary's office but reporting to him, and that was another element of the Taylor reforms which could not help but alienate Frederick Hutton. The man he picked was Morris Llewellyn Cooke, a Lehigh graduate in mechanical engineering and scientific management disciple. Cooke's first assignment was to overhaul the Society's printing and publication practises, a field in which he had several years of practical experience. It was that department Harrington Emerson had attacked for its wasteful methods at the Boston meeting in 1902, and Cooke soon claimed the changes he had installed would result in an estimated saving of 32 per cent. That kind of success seemed proof of the value of Taylor's campaign and the Council gave his Committee on Standardization, composed of the President, Secretary, and Fred J. Miller, a broad mandate that included the power to hire, fire, or reorganize the office staff.

Hutton had seen it all coming, even before it happened. The 1901 dues quarrel had revealed details of ASME's mismanagement that unavoidably reflected badly on its administrator, and Hutton must have known perfectly well from the attacks of Fred Halsey and Harrington Emerson that the scientific management crowd in the Society were eager to change its style and direction. It must also have been apparent that Taylor was headed for the presidency – occupants of that office were generally pretty well identified by the Society's insiders a few years before their election – and perhaps while Taylor was Vice President Hutton already had intimations of the new regime. Thus, even before his opponents brought in their program, Hutton decided the time had come for him to retire. In fact, he had entertained for some while the sentimental idea of resigning at the time of his twenty-fifth anniversary in the job.

But he was not without a feeling for the dramatic moment and instead he chose the very first Council meeting in Taylor's term of office to announce his resignation. The Society would soon be moving to the building Andrew Carnegie had donated to house America's national engineering societies and therefore, Hutton explained in a lugubrious letter to the Council,

I might emphasize the more obvious culmination of my service to the Society by withdrawing at the time when the Society shall cross the threshold of its new home in the Engineering Building. It will then have been given to me to have wrought for the Society from the days of small beginnings, when I paid my own office rent and expenses, through the period of rented offices and of the ownership of the modest home in Thirty-first Street and up to the opening of the days of larger opportunity in the splendid surroundings of the new enterprise.[20]

20 Council Minutes 30 Jan. 1906

Hutton's decision posed a political problem for Taylor on two counts. First, the Society's long-time Secretary had many friends in the organization and it was important that it not appear he was pushed out of the job. Second, some Council members, Jesse Smith, for example, were concerned that there was 'already too much routine' in the Society's operations, and Taylor did not want an upset over Hutton's resignation to derail his efficiency plan for the organization.[21] His solution was one the retiring secretary and his friends could hardly help but find appealing; Hutton was nominated as the Society's next president and he was to be named Honorary Secretary after that on the basis of a polite fiction that he would still perform some of the social functions of the secretary's office. What Taylor also realized was that his program depended upon the incoming secretary and he had a hand in solving that problem, too.

Several men prominent in New York technical journalism circles were mentioned for the post, including Lester French, editor of *Machinery*, and Henry H. Suplee, editor of *Engineering News*. But at almost the last moment the Committee on Standardization, to whom the responsibility had been delegated, settled on Calvin Winsor Rice of Winchester, Massachusetts, a thirty-eight year old electrical engineer then working as a consultant to the General Electric Company in New York. An 1890 graduate of Massachusetts Institute of Technology, Rice had been a member of ASME since 1900 and was appointed Assistant Secretary in June 1906, at an annual salary of $6,000, with the understanding that the Council would elect him secretary at the end of the year. What generally recommended him for the post was a reputation he had already acquired for what the profession liked to call engineering statesmanship. For example, along with Charles F. Scott of AIEE, Rice helped persuade Andrew Carnegie to donate the money for a new Engineering Societies Building, and organized the great American reception for Lord Kelvin in 1902. He had also taken a prominent role in the establishment of the John Fritz Medal, and these kinds of activities identified him with a set of altruistic, co-operative, professional society endeavors. Rice was also known to many in the Society through his effective work on the meetings committee, along with Charles Whiting Baker, another liberal engineer with reform tendencies. In addition, he must have been an attractive candidate to ASME's Council because of his engineering and administrative experience in electric power generation – a field that had assumed considerable importance in the Society.

Taylor had cleverly managed to give the Committee on Standardization, which Miller delighted in calling 'the iconoclastic committee,' the job of

21   Jesse M. Smith to F.W. Taylor, 10 Jan. 1906; Taylor Papers

picking the Secretary-designate and, since both of them controlled the committee, their candidate was certain to be one committed to efficient methods for the Society. Rice proved to have just that sort of enthusiasm and worked closely with Taylor and Morris L. Cooke to implement the reforms. Both Hutton and his own assistant, Francis Hoadley, found all these changes upsetting. As Rice tactfully put it, Hoadley was 'somewhat discouraged in the attempt to undertake the new system.'[22] Taylor was more candid. In a letter to Miller he revealed the tension between the two camps that Council minutes so completely masked: 'our friend Cooke has had a battle royal throughout this summer and fall with the Professor and Hoadley, and has done our Society a great service.'[23] But Taylor knew the completion of the work he had begun would rest with the next year's Council, and almost before Calvin Rice started Taylor was plotting to insure that the new Secretary would have a supportive group to continue with. The way to that end was through the nominating committee and Taylor assured Rice, 'I shall try to have my friends on the Nominating Committee appoint men who will be especially friendly to you.'[24]

Taylor's scientific management of ASME's business was first directed to its publications practices, since that was where most of the Society's income was spent. But the significant economies came from Cooke's experience in the printing industry rather from any management magic. He put the printing out to competitive bidding, bought paper and ink at more favorable prices, and organized publication processes to save time and materials. Cooke's new arrangements also increased effectiveness. Speaking of his own monumental paper, 'The Art of Cutting Metals,' printed in a special two-volume issue of *Transactions* as his presidential address, Taylor wrote to Rice, 'under the old management of the Society, however, it would have been entirely impossible to have gotten it out, and it is only due to the energy of yourself and your staff, and particularly Mr. Cooke, that it was printed at all.'[25]

Taylor's system was extended to practically every other aspect of the Society's business. For instance, Carl Barth, another disciple of scientific management, was brought in to design a whole new set of accounting forms. In just the same way that American industry during these years was also working out new accounting procedures to provide a much broader range of data,

22 Calvin W. Rice to F.W. Taylor, 9 June 1906; Taylor Papers
23 F.W. Taylor to Fred J. Miller, 8 Oct. 1906; Taylor Papers
24 F.W. Taylor to Calvin Rice, 7 June 1906; Taylor Papers
25 F.W. Taylor to Calvin Rice, 20 Dec. 1906; Taylor Papers

Barth's system aimed at a greater volume of information to give ASME's leaders better control over its financial affairs. And to avoid the kind of administrative break-down of 1901, office procedures were also completely standardized. Thus, the standardization committee reported to Council in the fall of 1907, 'the object of your Committee ... has been to organize the office and routine work of the Society that it would go forward largely independent of the Secretary.'[26] Its approach was a pure reflection of scientific management ideology and aimed at a functional organization of the Society, with each function in the hands of a single, well-qualified person, instead of the one-man operation Hutton had run so long. Besides, the reformers had in mind a more ambitious role for the Secretary. In their opinion, the Society should have a broader appeal and relevance for mechanical engineers throughout the country and they also imagined a wider public career for the organization itself. Thus, when the opportunity came to define the functions of the Secretary, they made it his job to lead ASME into that new era.

These hopes were shared by others in the engineering profession and the completion of Taylor's term of office provided an opportunity to express such feelings. *Engineering News* put its coverage of the 1906 annual meeting on the editorial page and entitled it 'A Forward Movement in the American Society of Mechanical Engineering.' Probably written by Charles Whiting Baker, and couched in the subtle language used to describe engineering politics, the editorial identified several major reform themes. First, the changes were popular. The Society always drew more attendance at its annual meetings than any of the other founder societies, but this one broke all records. The newspaper pointed out that while their 'social features' had invariably been a prominent part of ASME meetings, the quality of papers at this convention was 'far in advance of those at any previous meeting in a long time.'

These changes in the Society's fortunes, *Engineering News* made plain, had not come by chance – they were due to Taylor's energy and commitment. Most engineering society presidents took their election as an honor and lent what dignity they could to the office. But Taylor saw himself in the capacity of an executive leader and spent heavily of both his time and money to make an efficient and economical organization. Chief among his accomplishments was Calvin Rice's appointment. 'It is something of an innovation,' the editorial claimed, 'to select as secretary of an engineering society an engineer of

---

26 Council Minutes 12 Nov. 1907. See also Alfred D. Chandler jr *The Visible Hand: The Managerial Revolution in American Business* (Cambridge: Harvard University Press 1977) 109.

high standing in practical engineering work, and have him devote his entire time to the society's affairs.' But according to the newspaper, the high quality of the annual meeting's papers and discussion was 'largely due to Mr. Rice's initiative and energy.'[27]

Silently reflecting a long-standing criticism of Frederick Hutton, the *Engineering News* editorial elaborated on the importance of actively seeking good papers for meetings and the *Transactions*, since most members measured the return on their dues in terms of the value of the Society's publications. And in mechanical engineering, the editor argued, it was much more difficult to get good papers because business firms often refused to allow their engineers to make public the results of their investigations and practices. Accordingly, the Society's papers were 'too largely contributed by professors and dealt with theoretical matters or laboratory tests rather than with practical present-day work of the engineer.' Because of his own experience and standing in the technical community, however, *Engineering News* thought Rice 'preeminently fitted for the work of educating our great industrial corporations to the advantages of technical publicity.'[28]

Within the Society itself, and especially among members of the Council, Taylor's program did not fare so well. Even before his own presidential year was out he worried about Hutton's ability as his successor in office to dismantle his reforms, although Miller assured him there was little any president could do without Council support 'and I imagine that there will be ways of securing proper action by the Council.'[29] But it did not prove so easy to manage that body, and Taylor wrote to Rice 'it is a matter of regret to me, and I believe also to you, that the managers of the Society do not seem inclined to move in the direction of broadening the scope of its usefulness.'[30]

Not all of Taylor's reforms had proved as successful as Cooke's renovation of printing and publication. For example, it took a long time to sort out the accounting procedures, and extra bookkeepers had to be hired for the job. In fact, the elaborate nomenclature and Carl Barth's special forms cost money and Taylor was worried that his system might be associated with higher dues when the Council reacted to these increased expenditures. Not only that, it took more people to implement the changes and Rice reported to Taylor, 'we have nearly doubled the staff in the office over that of a year ago. I cannot see that with the completion of the introduction of the system we can get along

27 *Engineering News* (13 Dec. 1906) 620; also printed in *American Machinist* (1907) 150–1
28 Ibid
29 F.J. Miller to F.W. Taylor, 26 June 1906; Taylor Papers
30 F.W. Taylor to Calvin Rice, 8 July 1907; Taylor Papers

with any fewer people.'[31] On top of these extra expenses, the cost of printing Taylor's long paper on metal cutting technology ate up all the savings Cooke had achieved in the publication department!

One member of Council, Jesse Smith, was outspoken in his opposition to another implication of Taylor's efficiency plans. Smith objected to the managerial style the new system would impose and early in 1906 he explained to Taylor in a straightforward way: 'we all of us are hearty believers in the main work of the Society being done by the Standing Committees, rather than having that work done by a single person or by the paid employees of the Society.'[32] In time, that difference of opinion matured into open warfare and it is not difficult to imagine that part of Smith's opposition stemmed from the way the reformers went about things. Taylor had grown up in cultured surroundings and his correspondence reveals the polite care he took at the level of formal discourse. But in his zeal to impose his vision on the Society, he packed the Committee of Standardization and manipulated nominations in an obvious fashion. And Cooke, as ASME would soon learn, was not a man to mince words at all; instead of Calvin Rice's charitable assessment of Francis Hoadley, Cooke flatly argued the man was 'too stupid' to do the work assigned him.[33] Nonetheless, in mid-1908 Rice reported to Taylor that the reform program was going well. The accounting methods had finally been straightened out and two men favorable to scientific management, George Barrus and James M. Dodge, had been appointed to the nominations committee – 'to see to it that the personnel of the officers for the ensuing year will be in harmony with the structure for which you have laid the foundations.'[34]

The opposition was far from eliminated, however. In a move that strongly suggests a conservative reaction, the nominating committee proposed Jesse Smith for president in 1909. In consternation, Rice called the committee's attention to Smith's previous opposition to Taylor's reforms and, as he told the story to Taylor, the committee chairman, James Mapes Dodge, then secured from Smith a pledge to support the reform platform. Presumably on the strength of that promise, Smith was nominated and duly elected as Society president. But the understanding all came apart at the first Council meeting, of which Taylor was a statutory member as a past president. 'I had the riot act read to me by our new President just after the close of our last Coun-

31 Calvin Rice to F.W. Taylor, 20 May 1907; Taylor Papers
32 Jesse M. Smith to F.W. Taylor, 26 April 1906; Taylor Papers
33 F.W. Taylor to Fred J. Miller, 5 Jan. 1907; Taylor Papers
34 Calvin Rice to F.W. Taylor, 12 June 1908; Taylor Papers

cil Meeting,' Taylor wrote Rice. Smith's warning was that 'all progress and change in the Society must come either from the membership as a whole or from some of the Committees,' and he told Taylor in no uncertain terms that he would oppose any motion that did not originate from either of those sources.[35]

In his presidential address, Smith restated his ideas about Society governance. He stressed the importance of management by standing committees as well as the value of an active membership, and he emphasized that it should be increased 'only by men of high quality as engineers.'[36] In the discreet language the Society's leaders used in their public utterances to discuss what were really internal matters, Smith's words were euphemisms for a set of conservative policies. The members at large actually had few ways to influence anyone and, with the Society's focus so much in New York, even an array of standing committees still meant that a small group of men made the crucial decisions.

In fact, Taylor was not himself concerned to alter the Society's essential political features. Unlike his protégé Cooke, he was not that interested in the politics of engineering or in the democratic reform of the profession. Taylor was absorbed more with process than with purpose. Besides, there were limits to his own power. He worried about critics who claimed the Society was run 'by a clique of self-centered and interested men' and, he wrote to J.F. Klein in 1907, he wished to avoid even the suspicion that it might be true.[37] But he was forced to just such methods himself, in the same way he was compelled to appoint New Yorkers to important committees despite his conviction that the Society needed men 'from widely different sections of the country, and also of widely different interest and pursuits.'[38]

Yet Taylor's reforms coincided with a larger movement in engineering professionalism and pushed ASME to explore its responsibilities to its own members and to the American public. The Society's concern to eliminate administrative incompetence gave Taylor an opportunity to test the broader applicability of scientific management outside the field of industry, and there is no doubt that he seized the chance for personal as well as for altruistic reasons. His thirty-three – page set of standard instructions for the conduct of annual meetings was certainly in sharp contrast to previous practice.

35 Calvin Rice to F.W. Taylor, Tuesday [1908]; F.W. Taylor to Calvin Rice, 8 Dec. 1908; Taylor Papers
36 Frederick R. Hutton *A History of the American Society of Mechanical Engineers from 1880 to 1915* (New York: ASME 1915) 121
37 F.W. Taylor to J.F. Klein, 8 July 1907; Taylor Papers
38 Ibid

But there was more than minutiae in Taylor's methods. For example, he strongly believed in a link between his reforms and an increase of the membership, and he supported measures designed to foster geographic and technical subdivisions of the Society. He was on the committee that helped set up the Gas Power Section and his only objection to a 1907 constitutional amendment that hesitantly encouraged the creation of technical interest groups was that it did not seem 'radical enough.'[39] A larger membership, broadly representative of technical interests and geographic areas, also justified the additional costs of scientific management. Or, to put it differently, an effectively operated society could expect greater revenues. In just the way efficient workshop management led to increased production and larger profits, Taylor imagined that a more rational administration of ASME would allow it to realize its institutional potential.

From the beginning, scientific management suggested a new and enlarged field of action for engineers. Henry R. Towne's 1886 paper before ASME on 'The Engineer as an Economist' excited its listeners on at least two counts. It made management a legitimate professional interest for mechanical engineers, whose functions so often included the 'executive duties of organizing and superintending the operations of the industrial establishments,' as Towne put it.[40]

And, in mediating between the interests of capital and labor, management also pointed to a social role of vast importance. Towne's paper was presented in a special session on 'economic' matters that also included papers by Oberlin Smith and Henry Metcalfe. Most of those who took part in the discussion afterwards agreed that high wages for labor were to everyone's advantage; indeed, Charles H. Fitch argued that as a moral responsibility the Society 'ought to consider the condition of mechanic labor and the means by which it may enjoy a more gratifying compensation.' But engineers from shop backgrounds claimed it was 'the tendency of the average man ... to do as little as possible,' so it remained to find a mechanism that solved both problems.[41] Almost ten years later, Taylor's paper 'A Piece-Rate System, being a Step Toward the Partial Solution of the Labor Problem,' seemed to provide the answer. William Kent pronounced the subject 'one of the most important questions, not only before this society, but before the world to-day – the harmonizing of labor and capital.'[42]

39. F.W. Taylor to Jesse M. Smith, 22 April 1907; Taylor Papers
40 ASME Transactions VII (1886) 428
41 See ibid 469–88, for the discussion of all three papers.
42 Ibid XVI (1895) 891

Besides that exciting new dimension to ASME's social economy, as Edwin Layton has pointed out, scientific management also had appeal to younger members trying to define a professional status for themselves.[43] The Society's older leaders, men such as Towne and Oberlin Smith, managed industrial enterprises of their own and, while they had some interest in giving mechanical engineering a social and professional distinction it did not earlier have, at a personal level they already enjoyed the prestige America accorded the proprietors of large business firms. But the men who flocked in such numbers after 1900 to join engineering societies did not own factories – they worked in them; and although most of them shared the values and objectives of their employers, they sought an independent occupational identification. Taylorism, by its emphasis on applied science and social purpose, indicated one way these new bureaucrats of the urban industrial complex could still satisfy the sense of self-governing professionalism.

This vision of the profession's larger social purposes motivated Morris L. Cooke to shift his attention from ASME's administrative reorganization to the regeneration of its soul. His moral crusade began on a relatively modest note, but one much in keeping with the spirit of Progressive Era reform politics and Theodore Roosevelt's 1908 White House Conference of Governors that launched the Conservation Movement. In a paper presented at a 1908 annual meeting of the Society, entitled 'The Engineer and the People: A Plan for a Larger Measure of Cooperation between the Society and the General Public,' Cooke pointed out that of the three parties principally interested in engineering work – the engineer, his employer, and the public – only the last was usually ignored. But the time had come, he said, for engineers, 'as members of a public spirited profession,' to apply their specialized knowledge in the public interest 'with just as much fidelity and zeal as they work for their employer.'[44] Cooke pointed out that in none of the Society's many and varied activities were there any clearly dedicated to the public welfare – even though the constitution defined its functions broadly – and almost none of the thousands of papers read and published over the years were addressed to anyone other than engineers.

Instead of that neglect, Cooke proposed a broad range of programs that collectively aimed to inform the public about the engineering profession, to help voters understand political issues with technical content, and to aid government, at all levels, in framing policy. The heart of Cooke's plan was a

43 Edwin T. Layton jr *The Revolt of the Engineers: Social Responsibility and the American Engineering Profession* (Cleveland: Case Western Reserve University 1971) esp. ch. 6
44 ASME *Transactions* XXX (1908) 619

'Committee on Relations with the Public,' and lest his audience think such a concern for publicity unprofessional, he cleverly pointed out how much the medical profession was already doing along those lines. But he did not base his appeal only on altruistic grounds. 'It is public opinion and not the dictum of the engineering fraternity which finally decides the large questions of engineering practice,' Cooke claimed, and he argued that if the profession were to grow in the public's estimation, it needed to do a better job of public relations.[45]

As an example of the sort of technical problem in which the public's interest was obvious, Cooke put together a plan in 1909 for a national conference of air pollution, especially that from industrial smoke-stacks. Rather than a conventional technical session, his conference would feature the judgments of architects and physicians as well as of engineers, and it was calculated to attract public notice. In fact, Cooke had not imagined that the conference papers would be printed in the Society's *Transactions* at all, and aimed instead at a format that would attract newspaper attention. He knew there would be opposition to his plan, that besides its unprofessional style it also touched sensitive nerves, so he prepared his appeal with considerable care. He framed a petition on behalf of his conference and then confidentially circulated it among a number of members, including 'some of our most competent authorities on smoke' and other men 'closely connected with the present management of the Society,' before sending it out to a still larger number of members for their support. But despite the fact that the signers included fifteen Council members, a number of past presidents and former members of the meetings committee, prominent professors of mechanical engineering, and the chief engineers from 'a number of companies operating large power plants,' the committee on meetings, chaired by W.E. Hall, a New York consulting engineer, peremptorily ruled against Cooke's conference.[46]

ASME's Council minutes report the episode with little detail: 'The Meetings Committee reported that they had declined the petition of a committee consisting of Messrs. Charles H. Benjamin, James M. Dodge, J.J. Flather, H.V.O. Coes, and Morris L. Cooke and about 150 members of the Society for a National Conference on the Subject of Smoke Abatement.'[47] And with even less information about its own debate on the issue, the minutes record

---

45 Ibid 627
46 Willis E. Hall to Morris L. Cooke, 13 May 1909; Cooke Papers, Franklin D. Roosevelt Library, Hyde Park, NY
47 Council Minutes 2 June 1909

only that the Council approved the action of the meetings committee. But in a letter he wrote to William H. Bryan of St Louis, a member of the committee who had been unable to attend when the matter was decided, Morris L. Cooke outlined what he thought had happened: 'I think it is a fair statement to say that the only substantial opposition to the petition originated south of 14th Street in New York City. Since the Public Service Commission was appointed some years ago this question of smoke has been given more thought in New York than before and unfortunately some people looked upon our petition as having too close an application to this recent bit of New York City history. Mr. Scott of the Interborough Co., Mr. Lieb of the Edison Co. and Mr. Whyte of the New York Central R.R. all doubt the advisability of any action along the lines suggested.'[48]

There was no single particular argument against the idea, Cooke told Bryan. The committee chairman thought an investigation into 'the ultimate nature' of flue gasses would prove a more effective approach to the problem and that position was supported by Professor David S. Jacobus of Stevens Institute, who had served the New York Edison Company as its expert before the Public Service Commission. Others opposed the conference because of its popular quality, because it threatened the possibility of 'unbridled speech,' or because it would not be 'strictly engineering' since it involved non-technical participants. Most disappointing of all to him, Cooke said, was Calvin Rice's reaction, and he quoted part of a letter the Secretary had written to him: 'Such men (Mr. Lieb, V.P. New York Edison Co. and Mr. Stott, Supt. of Plants of the Interborough Co.) *and only such men* [italics mine] should have a prominent position in relation to the committee organising such a conference so as to guard the Society against unfortunate and detrimental action to vested interests.'[49]

What appears an obvious conspiracy of private utility engineers was something more complex than that, however. As time would make clear, the utilities and their suppliers constituted a powerful force within ASME, but they did not automatically unite on all issues touching their interests. For example, Cooke's petition was signed by, among others, Alex Dow, vice president and general manager of Detroit Edison, Alexander C. Humphreys, president of Stevens Institute and a consultant to the gas industry, Irving E. Moultrop of Boston Edison, Professor Mortimer E. Cooley of the University of Michigan, who also consulted for electric utility companies, and such conservative Society stalwarts as Henry R. Towne and Frederick R. Hutton.

48   Morris L. Cooke to William H. Bryan, 11 Oct. 1909; Cooke Papers
49   Ibid

The group that killed Cooke's conference was made up of New Yorkers. They did it for essentially local reasons and they were able to do it because men from the city were over-represented in ASME affairs. As Frederick W. Taylor had realized, the only way important committees could get their work done was if members within easy travelling distance made up the necessary quorum. In return for this heavier burden of service, they reaped disproportionate power within those committees. The Council was supposed to be a more broadly representative group, but even there geography still worked in favor of New York area members. For instance, at the session which approved the meetings committee decision on the conference, only nine members of the Council were present, plus the President, Jesse M. Smith, and Calvin Rice. An equal number of Council members had signed Cooke's petition but were not present and for some of them distance had been a factor.

Air pollution was the perfect Progressive Era issue. It suggested the inefficient consumption of fuel and the waste of natural resources; it also implied the waste of human resources through consequent ill health, and of financial resources to pay for the clean-up. Furthermore, the control of industrial pollutants called for expertise and balanced judgment since regulation in the public interest included a concern to protect investment and employment. Cooke could hardly have picked a better subject to illustrate the way a public-spirited profession might win esteem and carry out its social responsibilities to the nation. It was also precisely the kind of issue municipal engineering societies became involved in and for exactly the same reasons. Indeed, the difference between the enthusiasm with which local groups grasped these opportunities and ASME's reluctance to deal with them illuminates another important aspect of the Society's social economy and suggests once again how questions of its governance and purpose were tangled with geographic concerns.

Municipal engineering societies proved such immediately useful vehicles for reform because they tended to be more sensitive to the career interests of younger men. They had lower admission requirements and lower dues than the national societies and they more actively promoted campaigns to improve the economic and social status of younger men. Just as Cooke had proposed, these local associations encouraged a public-spirited approach to engineering on the grounds that an enhanced estimation of the profession would lead to more use of it in technical matters of public interest. The history of the Cleveland Engineering Society during the first two decades of the twentieth century provides a good example of how the Progressive reform movement and the professional aspirations of younger engineers reinforced each other.

In 1901 Cleveland voters elected Tom L. Johnson mayor on the strength of his promise to bring about a broad set of reforms that included street paving and lighting, bridge building, sewer construction, new waterworks facilities, and the expansion of the city's public utilities – the kind of municipal engineering, in other words, that most American urban centers needed at the turn of the century to cope with massive increases in their populations and with the impact of a similar scale of industrial development. The local engineering society, established in 1880, the same year as ASME was founded, soon became caught up in the city's reform mood and in 1904, for instance, appointed a committee to study water pollution. In its report the following year, the committee identified the major sources of industrial pollution, particularly that of the Standard Oil Company, and proposed a set of legislative measures to correct the problem. Over the next several years the Cleveland Engineering Society had committees investigating such topics as smoke abatement, boiler inspection, and building construction standards, and all from a standpoint of professional involvement in public affairs explicitly contrasted with the conservatism of the national societies.[50]

It was an article of faith in municipal societies such as Cleveland's that engineers did not enjoy the same esteem as doctors and lawyers. ASME's leaders were not much troubled by those kinds of comparisons, but to younger men in an increasingly bureaucratized field, professional status had become important. Objective and expert service to the community was the simple and direct way to improve their status. Furthermore, they argued, by demonstrating their interest in public issues engineers not only gained respect, but also increased employment opportunities – something else that concerned rank and file members. In Cleveland, therefore, engineers linked professionalism with public service. And because municipal societies brought local engineers together regardless of their disciplinary interests, those organizations tended to think of the profession the same way when they sought mechanisms for political or social action.

The movement among urban engineering societies to redefine the social economy of the profession so as to encompass its responsibility to the public coincided exactly with Morris L. Cooke's crusade in ASME toward the same goal. These separate but related developments shared language, methods,

50 Bruce Sinclair 'The Cleveland "Radicals": Urban Engineers in the Progressive Era, 1901–1917' unpublished research paper, 1965

and actors. As they unfolded, however, one forced the Society to work out a new set of relations with its membership, while the other compelled it to reconsider Thurston's claim for the identity of the private and public interest. Neither process came easily, since at the same time urban and national societies were also engaged in a contest for institutional hegemony within the engineering profession.

Cooke's paper 'The Engineer and the People,' which proposed an activist role for the Society in public affairs, was well received by those who heard it and mentioned it in the technical press. In fact, it did not present very radical ideas. Arthur T. Hadley, president of Yale, and principal speaker at the dedication of the new Engineering Societies Building in April 1907, had called upon engineers to commit themselves to public service in much the same way Cooke did. But Cooke's efforts to put those principles into action stirred up strong opposition and led him into a confrontation with leading figures in ASME over the most sensitive of issues – the use of specialized knowledge in a democracy.

Even in the exuberant moments of Taylor's presidency, when the scientific management enthusiasts of the Society looked forward to its complete renovation, Cooke must have known that the organization was deeply conservative in its tendencies. But if he did not, there were soon plenty of examples to instruct him. Jesse Smith's term as president signalled the beginning of a series of reactions to Taylor and his movement. Besides Smith's own vendetta, which forced Taylor from the Council's meetings, the Society refused in 1910 to publish his *Principles of Scientific Management*, and two years later a special committee rejected his claims for the scientific basis of his management system. As those measures drove Taylor still further away from ASME, Cooke's experiences drew him closer in a fervent wish to win the organization to the side of public service. But after 1912 Cooke's strength of conviction flowed more from moral outrage than scientific management; as director of Philadelphia's public works in a Progressive reform government of the city he had discovered at first hand how much graft and profiteering existed in municipal engineering. Cooke was particularly offended to learn that while the utility companies were able to command the services of the country's most prominent technical men, the city was hard pressed to find accurate and objective engineering advice. That lesson led him in 1914 to organize the session for ASME's annual meeting that again aimed to preach to his fellow engineers the gospel of civic consciousness.

At that point, Cooke's ambitions intersected with those of Cleveland engineers bent on a reform program of their own. As an essential weapon in its campaign to inform the community about engineers and engineering, the

Cleveland Engineering Society had appointed a publicity committee in 1912 and named C.E. Drayer, a young assistant engineer, as its chairman. Drayer soon proved an adept at popular writing as well as a zealous publicist and he eagerly went from a successful series of articles for the local newspaper on 'Engineering as a Life Work,' to instructive pieces in the technical press, and then on to what he called 'missionary' labor among other engineering societies. His activities stirred up a great deal of interest and when Cooke came to put together his session on public service, he got Drayer to present a paper entitled 'The Engineer and Publicity.' Cooke planned to offer a paper of his own that he called 'Some Factors in Municipal Engineering,' and for one of the technical papers he went back to another Cleveland activist, Frederick W. Ballard, the city engineer responsible for the design and construction of the city's new municipally owned electric generating plant, who agreed to speak on that subject.

The session had been described, innocuously enough, as an attempt to show engineers the needs and opportunities for their skills in civic affairs, but Cooke had another motive as well. Cleveland's municipal power plant was profitably selling electricity to its customers at less than one-third of the price the city's major private utiltiy claimed it needed to charge in order to meet its costs and pay a return on investment, and Cooke wanted that comparison driven home. He knew from experience, however, that his proposal would never be accepted as a technical session so he adroitly brought it in under the wing of the public relations committee and advertised it as a public service meeting.

Morris L. Cooke's relations with ASME form one of its most fascinating chapters. He upset its leaders more and longer than anyone, but better than anybody else he identified the ills inherent in the Society and reflected some of the profession's highest ideals. To its benefit, although in a manner the Society's officers could hardly appreciate, Cooke forced the organization to deal with the most pressing problems of its social economy, which in 1914 consisted of geographic balance, democratic structure, and professional responsibility. All those issues were involved in the public service session and that was when his long battle with the Society began in earnest.

While it dealt with subjects unfamiliar to ASME meetings, the tone of Cooke's paper 'Some Factors in Municipal Engineering' was relatively straightforward. He meant to convince the Society's members that America's cities needed their technical judgment and that public service was an honorable calling. That last part was especially difficult since municipal politics was so often corrupt that most people assumed city engineers to be 'political placemen' or of second-rate ability. As one of Cooke's listeners reminded him, James Bryce's *American Commonwealth* had long ago pointed out that the best men do not go into politics. But Cooke linked the reform philosophy of the Progressive Era and the ambitions of many younger engineers into a logical and persuasive argument. He began by noting that the growth and concentration of corporate power had tended to absorb engineering skills to such a degree that cities found themselves unable to acquire objective technical information. In purchases, whether of asphalt, or concrete, or electricity, municipalities discovered that the most prominent men in the field were already on retainers from a supplier, so 'that for the average city official to get good advice on these matters is well nigh impossible'[51]

The remedy to that situation, according to Cooke, was for engineers to make a whole-hearted commitment to the public's interest. And he claimed that the public was 'ready to accord the engineer a leading, perhaps controlling part' in the community if he stood willing to give it his complete devotion. Carrying a comparison with the legal profession much farther than engineers usually did, Cooke said that as a scientist responsible for facts regardless of their advantage to anyone, 'the service of an engineer ought to be as the service of a judge.' In fact, he put a great deal of emphasis on the necessity for a civic-minded viewpoint, in contrast to the obligations engineers in the private sector felt to their employers, and he claimed that from a maturing sense of broader responsibilities, engineering had risen to 'the stage of development where it has become a profession in the highest sense of the word.'[52]

It still remained, however, to convince the public that engineering skills were crucial to the efficient management of America's cities and that it was worth spending tax dollars to make municipal service an attractive and remunerative career. Cooke told his audience it was up to engineers themselves to bring about that shift in public opinion and he said the way to do it was by advertising. What he meant was an active campaign of public education to teach people the importance of expert knowledge – and that implied taking the mystery out of it. Finally, by way of example, Cooke suggested the national societies might act as a sort of civil service board to help cities find the best engineering talent available, and a separate section of the engineering societies' libraries be designated a municipal reference library, to create the resources for this new science of municipal engineering.

The element of Cooke's paper which most offended was his argument that a separate, civic, viewpoint was needed in the profession; and those who

51 Morris L. Cooke 'Some Factors in Municipal Engineering' ASME *Journal* (Feb. 1915) 82
52 Ibid

reacted the strongest were engineers connected to utilities. Alexander C. Humphreys, president of Stevens Institute and a prominent consultant to the gas industry, angrily denounced 'the pervading tone of the paper, which seemed to imply that because an engineer has been in the service of the public utilities he is not to be relied upon to give honest advice in connection with public affairs.' Anyone making that sort of accusation, he said, was 'unworthy of a place in the profession.'[53] Alex Dow of Detroit Edison also rejected the implication that engineers who had worked for the utilities were somehow tainted by that association, and the rest of the discussion largely came down on one side or the other of that issue. Harlow Persons, of the Amos Tuck School at Dartmouth and an active figure in scientific management circles, spoke for Cooke's position when he said that engineers needed educating to get them to exchange 'the motive of private gain' for 'the motive of public service.' And Charles Whiting Baker noted that if engineers developed a sense of social responsibility, the new city manager form of government opened up the kind of important role in civic affairs for them that Cooke's paper had claimed was possible.

His friends were quick to point out that Cooke had not directly accused anyone of being disqualified for public service because of previous work for private utility companies. But there was no doubt in anyone's mind he meant just exactly that. From his experience as director of Philadelphia's public works, Cooke had come to perceive the privately owned utility comapnies as the city's natural enemies. And from his participation in ASME affairs, he had come to the conclusion that utility company engineers exercised a disproportionate influence in its government and that they used their power to subvert the Society's social responsibilities. Thus, while the sounds of battle soon became more strident, the essential lines of it were drawn at that 1914 public service meeting.

The utilities provided Cooke with an ideal foe, for several reasons. They were popularly identified with inscrupulous financiers and shady stock manipulation. Municipal reform campaigns of the Progressive Era had already drawn attention to the unseemly relations that so often existed between city politicians and utility companies, while at the state and national levels of government the principal thrust of Progressive legislation was the regulation of public utilities and of monopolistic practices. Cooke thus had the benefit of a general suspicion of utilities. But the juxtaposition of utility company engineers and an objective civic-mindedness had an especially sharp edge for ASME. The utilities were heavy consumers of mechanical engineering skills.

53 See ibid 85–7 for the discussion of Cooke's paper.

Electric power generation in particular depended on steam-engines and steam-boilers and their manufacture made up two of the most important fields of mechanical engineering. The utilities and their suppliers were also among the largest employers of mechanical engineers and those industries were the Society's strongest supporters. Their engineers filled the organization's highest offices and Cooke's indirect but obvious accusation precipitated a drawn-out and acrimonious quarrel.

Most immediately, the conflict took the form of a disputed election for a seat on the Council. Cooke had been nominated in 1914 for a one-year term as manager, an office usually considered the stepping-stone to a vice presidency, but his election was strongly opposed by utilities-connected members of the Council, apparently led by Alexander Humphreys. Through the hard work of a reform-oriented group in the Society, Cooke won the election and his very first action was to undertake an investigation into the utilities' representation on Council and the nominating committee. It seemed to him as if the utilities wielded the real political power in the organization and he began to campaign to open up the nominations process, particularly so that men who actively supported a 'public' position on utilities might be elected to office.

As a dissenter, Cooke was concerned with ASME's defensiveness in the face of criticism. Dissent was 'a sign of life,' he told Calvin Rice, 'and not to be deprecated.'[54] But he also saw it as a creative element in the democratic process of information-sharing that ought to go on between the Society and its members. Indeed, the control of information – whether of technical data or about ASME business – was at the heart of his crusade. Just as C.E. Drayer had inaugurated a campaign to publicize the work of Cleveland engineers, Cooke seized upon publicity as the essential weapon in the reform of the Society and in his war with the utilities. But his program of publicizing the problems of municipal engineering very quickly escalated the war between him and his colleagues who worked for the utilities. In the spring of 1915 he delivered lectures at several universities on 'the changing attitude of American cities toward the utilities problem.' One of those talks was at Harvard and in it Cooke criticized by name Alexander Humphreys, Mortimer E. Cooley, Dugald C. Jackson of Massachusetts Institute of Technology, and George F. Swain of Harvard for their collusion with the utilities to keep rates at an unnecessarily high level. He could hardly have chosen more formidable opponents and they immediately began a movement within ASME's Council to censure him for unethical conduct.

54   Morris L. Cooke to Calvin Rice, 22 June 1915; Cooke Papers

At about the same time, Cooke published his lectures in pamphlet form under the title *Snapping Cords* (which meant to suggest the movement of cities to break the bonds with which the utilities had made them prisoner) and he circulated it widely to engineers, city officials, and even the clients of the men he had named.[55] His opponents struck back through the nominating committee, which in 1915 had on it Edwin M. Herr, president of Westinghouse, as chairman; Richard H. Rice, of General Electric; Edward C. Jones, Chief Engineer of the Gas Department of the Pacific Gas and Electric Company; Wallace M. McFarland, of Babcock and Wilcox; and Fred J. Miller, factory manager of Remington Typewriter Company – and the only person on the committee not connected with the utilities. For president of the Society they proposed David S. Jacobus, of Stevens Institute, who consulted on an almost full-time basis for Babcock and Wilcox, a principal supplier of boilers to the utiltity industry. Jacobus was duly elected and one of his first acts was to dismantle Cooke's public relations committee.

The censure movement extended into Jacobus' term of office, but finally amounted to little because Cooke threatened legal action since he had been denied a hearing before the Council. Meanwhile, he began a campaign to have the National Electric Light Association ejected from the Engineering Societies Building on the grounds that its essentially commercial nature contravened the tax-exempt status of a building that housed non-profit educational institutions. And in 1917, in his continuing effort to publicize the way the Society worked, Cooke issued another pamphlet, called *How About It?*, in which he attacked ASME's secrecy, its undemocratic procedures, and the control exercised over it by New York big business. The pamphlet was broadly aimed at ASME's rank and file members and his principal argument was that the inefficient form of its government – in which the Council had authority but no time, the secretary continuity but no power, and the president only status – allowed a small group of New Yorkers in the executive committee, who represented mainly utility interest, to control the Society. And not only was its structure wrong, Cooke argued, it was fundamentally undemocratic. Instead of publicizing its deliberations, Council edited its proceedings to eliminate the discussion surrounding important issues. The minutes never revealed who voted or how. 'Our 7,000 members are a peculiarly red-blooded body of men,' Cooke claimed; 'I don't believe that it would cause the Society much trepidation to actually know something about what goes on in the Council.'[56]

55 See Layton *Revolt* 154–72 for further information on Cooke's fight with the utilities.
56 *How About It?* (Philadelphia: privately printed 1917) 7

Because he was an articulate and passionate reformer who dealt in ideals, Cooke is an attractive historical figure. And because he preserved his correspondence, while Calvin Rice and Alexander Humphreys did not, his own view of events gets imposed on the historical record. But he over-simplified the relation of the engineer to his employer. The Society's members were not scientists, as Cooke had suggested, concerned only with facts and not with the advantages that specialized knowledge might yield. But it was not the case that they were only economic men, untouched by intellectual challenge or social implication, or that they always acted with the fixity of purpose that characterized so many of Cooke's actions.

In 1917, for example, Cooke became involved in a power plant symposium jointly planned by the Boston sections of ASME and AIEE. The secretary of the local ASME section, W.G. Starkweather, a district sales manager for C.H. Wheeler Manufacturing Company, wrote Cooke in February inviting him to 'take a leading part' in the meeting and promised that because of recent hearings into Boston Edison Company stock manipulations, the session would be especially interesting and that its object was to consider municipal and independent plants as well as those of the large utility conglomerates. F.W. Ballard was also to be invited, since his Cleveland plant had proved such a successful example of municipal operation. In his reply, Cooke was concerned to know if the side of public ownership was to be equally represented on the program and he wanted to be sure Starkweather understood that he was 'definitely on the public side of all utility questions.'[57]

Starkweather's answer was enthusiastic. He assured Cooke that the program planners were in sympathy with his ideas 'to the limit,' and implied that a few of Cooke's 'bombs' might aid the local cause for a municipal plant. He was 'an humble disciple of individualism' Starkweather told Cooke, and he said, in language one does not usually associate with district sales managers, 'my flag is nailed to the mast, and I care very little as to whether the bloated capitalist likes it or not.'[58] As the date for the meeting, 5–7 April 1917, drew nearer, however, the Boston Edison Company apparently put pressure on the local organizers to change the format, and in reply to Cooke's question about it Starkweather admitted, 'it is true the Edison people forced a readjustment of our plans, with the addition of a lot of power plant details which crowded the original program.'[59] In the end, the sympo-

57 W.G. Starkweather to Morris L. Cooke, 12 Feb. 1917; Morris L. Cooke to W.G. Starkweather, 15 Feb. 1917; Cooke Papers
58 W.G. Starkweather to Morris L. Cooke, 17 Feb. 1917; Cooke Papers
59 W.G. Starkweather to Morris L. Cooke, 26 March 1917; Cooke Papers

sium's program was printed without either Cooke's or Ballard's name on it and neither went to Boston for the meeting.

There seems to be no question that Cooke and Ballard were 'disinvited' because of pressure from the Boston Edison Company. Starkweather said so and Fred Low, editor of *Power*, claimed that he had seen the letter from Charles L. Edgar, an ASME member and president of Boston Edison, protesting against the discussion of municipally owned generating plants. Low ran a sarcastic editorial on the affair and, making the same point in a letter to Cooke, he said 'it is a question whether the policy of a great national professional society is to be determined by the Public Utility interests.'[60] But while it is clear Cooke and Ballard were excluded because of their public interest bias, it is less certain that the Society was as threatened by the private utilities as Low claimed; he was elected vice president in 1919 and president in 1924. Further, the utility engineers were not as one-dimensional as Cooke painted them. Irving E. Moultrop was employed by Boston Edison and he presented a technical paper at the power plant symposium. But he was also one of the most energetic membership chairmen the Society ever had and, besides encouraging younger members, he actively promoted the development of local sections on the grounds that they made for a more democratic Society – exactly the remedy Cooke proposed.

In his single-minded pursuit of the utilities, Cooke made life difficult for his friends as well as his foes. Charles Whiting Baker, a staunch supporter and long-time champion of democratic measures in ASME, found even his private correspondence to Cooke published in *How About It?*, as the latter relentlessly maintained that everything spoken or written which touched the profession should be open to public discussion. In fact, Cooke had so angered some members, his friend Fred J. Miller pointed out to him, that they would fight anything he proposed, 'regardless of its merits.'[61] Still, the odds are that Cooke's continual agitation moved ASME in the direction of liberal reforms. Or perhaps the harmony between his ideas and those of municipal engineering societies such as Cleveland's was the combination that finally forced the Society to a revision of its nomination procedures, to an active policy of encouraging local section development, and to a wider view of its social responsibilities. There is some basis for the second proposition. Cooke's battle with the utilities in ASME coincided with the movement among city engineering societies to define a broad social role for the profes-

60 'The Boston Symposium' *Power* 17 April 1917; Fred R. Low to Morris L. Cooke, 27 March 1917, Cooke Papers
61 Fred J. Miller to Morris L. Cooke, 14 June 1920; Cooke Papers

sion. And the co-operative efforts of local groups to create alternative national institutions were a powerful incentive to the founder societies to pay more attention to the social and economic needs of young engineers, who after 1912 began to come into the profession in huge numbers.

The creation of these alternative organizations helped force the Society to redefine its social purpose, but the new groups, of which the American Association of Engineers is the best example, were called into existence because national societies seemed to concentrate so much on technical purpose. Founded in Chicago in 1915 as a direct response to the lack of engineering jobs, AAE consciously aimed to fill an institutional vacuum in the profession. At its first meeting, Ernest McCullough forcefully expressed the needs that younger engineers felt ASME and the other founder societies were not satisfying. It might be fine for those established associations to consider themselves strictly educational and thus demonstrate the standing of their members, he said. What young engineers wanted to know from their elders, however, was not technical details they could read for themselves but 'how to obtain a position and how to hold it.' Reflecting the trade union sentiments of some of AAE's founders, McCullough claimed there were 'thousands of young fellows who are earnest and are not yet anarchists or IWW propagandists,' but he obviously implied that without some help that is what they would become.[62]

Frederick Haynes Newell, former chief of the United States Reclamation Service, and an ardent conservationist, also spoke at the initial meeting of AAE, but painted a different sort of picture and suggested a different solution to the problem. He pointed out that less than one-quarter of the country's engineers belonged to the national organizations, an indication that the great majority of the profession was not 'fully employed to the best advantages to society and to themselves, with corresponding loss to all.'[63] To correct that sort of waste and inefficiency he argued the need for new co-operative arrangements among engineering associations. Newell agreed that the 'material welfare' of engineers was the profession's central problem but he sketched out a peaceful way to solve it that directly echoed Cooke's ideas and those popular among local engineering groups. First, engineers must have a concept of ethical public service. Newell drew upon the conservation movement for an example of the altruistic spirit he meant, and described an expanded sort of

62 'Time at Hand When the Engineering Society Should Awake to its Deficiencies' *Engineering Record* (2 Oct. 1915) 421–2
63 'Awakening the Engineer to Service Through Civic Responsibility' *Engineering Record* (2 Oct. 1915) 421

resource management that encompassed 'every improvement in the conditions favorable to human health, comfort and industry.' Second, it was incumbent upon engineers to 'sell' themselves to the public and he claimed that a combination of publicity and altruism would lead to full employment and enhanced professional standing.

Newell had previously taken his ideas of local engineering society cooperation to ASME's spring meeting in 1915. In a talk entitled 'The Engineer as a Citizen,' he reiterated the theme that greater public service would lead to greater public recognition for the profession. But, he argued, 'My personal belief is that the engineering profession as a whole is capable of being more immediately advanced in public esteem through strong, active, local societies than through any other one agency.'[64] The message was one a national society could hardly find attractive and Newell turned instead to local groups for a means to unify all engineers in a movement that applied technical expertise to public service in the interest of job security and conservation ideals. The American Association of Engineers proved the institutional mechanism needed and, together with C.E. Drayer, the Cleveland Engineering Society's publicist, Newell led the organization to a membership of over 20,000 by 1920. Although he opted for a different institutional format than Cooke, Newell's campaign was just as important to ASME because it helped define a set of member interests that after 1920 became an integral part of its social economy.

Cooke and Newell were thus spokesmen for a powerful set of forces in engineering professionalism during the first quarter of the twentieth century. To the reformers of ASME, scientific management had particular appeal because it promised a new kind of professionalism for engineers submerged in industrial bureaucracy and because it opened onto a role of public importance through the application of efficiency techniques to a wide variety of human activities, the most important of which was in the solution of labor disputes. But Taylorism was not the only source of reform inspiration. The Progressive Era's concern to bring order to America's urban industrial complex depended upon engineering skills and infused them with a sense of mission. And conservation, the heart of Progressivism, was ideally suited to be an engineering cause since it called for the efficient use of resources, made clear the need for experts, and promised them an important political place in a nation to be regulated by professional knowledge.

These ideas found different expression in different organizations. The Cleveland Engineering Society pursued a career of public service in the hope

64 F.H. Newell 'The Engineer as a Citizen' ASME *Journal* (July 1915) 6

of greater esteem for the professional status of engineers, better appreciation of the ways in which engineers would solve the city's problems, and thus more employment opportunities for its members. And to give its sentiments the weight of a national movement, it joined Frederick H. Newell's Conference on Engineering Cooperation, becoming, in effect, the first chapter in a national federation of local societies under the aegis of the Americal Association of Engineers. AAE, in turn, moved on to efforts to frame a code of ethics for the profession, secure the passage of licensing laws, formulate salary scales for engineers, and undertake related activities on behalf of the profession's material welfare.

Within ASME, this complex of reform ideas also produced a set of effects that substantially shaped its subsequent history. Taylor's reform of the Society's headquarters operation and its simultaneous move into the Engineering Societies Building made it a business office instead of the comfortable, club-like place it had been, to the regret of some members. That same application of efficiency methods also fastened a permanent staff bureaucracy on the organization, with an obviously important (although largely unrecorded) role to play in its future. The reform era reshaped some of the Society's election procedures. The pressure of distant members and Cooke's agitation forced the Council to give local sections' delegates, representing five regional groupings, the power over nominations. That action established the concept of regions and elaborated the Society's geographic structure; it also further enhanced the emerging political power of the Committee on Local Sections.

The growth of sections was seen as another reflection of more democratic tendencies within ASME and as an acceptance of a wider social role for the Society. While formal impediments to the establishment of sections were largely removed by 1909, it was not until Irving Moultrop's Committee on Increase of Membership was formed in 1912 that the number of new members each year provided enough impetus to overcome old fears that section activity threatened headquarters control and a lower membership standard. But in the five-year period between 1915, when the Committee on Local Sections was established, and 1920 ASME's membership doubled to about 13,500, while section membership almost tripled to about 8,500. An important element in that dramatic increase was the Society's accommodation to the fact that affiliation with local engineering organizations was the best way not only to keep the loyalty of distant members, but also to generate new ones. There is little in the Council minutes which describes the evolution of that point of view, but by 1919 it was firmly enshrined in the report of Louis C. Marburg's Committee on Aims and Organization.

Indeed, the recommendations of Marburg's committee gave an official cast to many of the ideas that had emerged over the previous two decades. The report urged an expanded employment service and a revised code of ethics for the Society. It proposed an active role of public service in such fields as safety, environmental protection, and the formulation of standards for consumer goods. There were proposals to aid the establishment and operation of municipal utilities, to help cities find engineers, and to publicize the profession of engineering. But perhaps the most striking feature of Marburg's report was its enthusiasm for the unification of engineering societies through 'some agency of a national character,' strong and independent, yet speaking and acting for the entire profession in matters of public interest.[65]

When the swelling movement for engineering involvement in public affairs was just gathering momentum, the frequent complaint of younger engineers from St Louis, Cleveland, or Chicago was that ASME and the other founder societies were too limited in their social purposes. What they meant was that the national groups seemed to be elitist and conservative, occupied mostly with technical papers and a style of meeting designed to reflect on their member's standing in the profession. Some critics claimed that the quality of papers and discussion was also pretty limited, but the main objection was that the national societies were not vitally engaged either in the interests of rank and file members or in the great national issues of the day. By 1920, neither of those charges could have been fairly sustained against ASME. The reform era had produced a remarkable variety of changes in practically every aspect of its social economy, defining its essential nature for the next several decades.

Against this appearance of dramatic difference, however, some fundamental things remained the same. The terms of employment for most of the Society's members were unchanged; despite the attention given to public affairs, most members still worked in industrial bureaucracies. The geographic balance of the Society had not changed – the great increase in membership had not by 1920 significantly altered the percentage of members based in New York. And notwithstanding new nominations procedures, in practice a small group of men in ASME still controlled the knowledge of its workings and made the most important decisions about its policies.

In the analysis of an individual life, biographical details tend to exemplify the subjects's essential characteristics, those elements of personality that remain the same in the face of change. The biography of an institution is like that, too. There are persistent elements which are continuous through its history and this is what makes an episode of one era seem so much to parallel another. Thus, the reform turmoil of the early twentieth century, which raised issues that have such current relevance, likewise reveals certain essential historical elements of ASME's social economy. And not only would questions about the organization's governance, geographic representation, and public responsibility continue to prove centrally important down through the years, but the way in which they were discussed would also have a familiar quality.

For example, although most ASME members of that period were politically conservative and tended to reflect a business point of view, many also gave the Society their time and energy for unselfish reasons. Therefore, more important than the verification of Cooke's charges is the recognition that inherent in the organization were both the capacity for the abuse of specialized knowledge and the capacity for altruistic public service, and that an important aspect of the Society's history has been the conflict between those two tendencies.

4

The Technical Sphere of Action

ASME's technical function, that other main stream of purpose dominating its history, seems at first glance very different from the shifting congeries of ideas and projects which made up its social economy. If nothing else, the technical objectives were straightforward and always remained essentially the same; at least they were described that way – to promote mechanical engineering knowledge and its useful applications. The men deeply involved in them have often thought the Society's technical activities its true work, the real basis for its claim to professional status and practical importance. And the ease with which ASME moved into the creation of standards, for example, seemed to indicate not only a natural role for the organization, but one of great value. However, the attempt to create a structure representative of the varied technical fields of interest within mechanical engineering, equivalent to that of the Society's geographic units, revealed more similarities than differences between the social and technical spheres of action.

At the Society's beginning, its founders took some pleasure in pointing out how many kinds of human endeavor were touched by the various and far-reaching branches of mechanical engineering, and that theme was repeated in more than one presidential address in the years that followed. It was also assumed from the start that members might logically be grouped according to their particular interests. After Henry R. Towne's 1886 paper 'The Engineer as Economist,' for instance, there was considerable support for the creation of an 'economics section' of the Society. None was actually formed – partly because so many members were interested in business management it would have been a 'committee of the whole,' as one pointed out; but the idea that ASME should give some scope to the specialties of the field had at least found expression.[1] In fact, however, the organization proved just

1 ASME *Transactions* VII (1886) 477

as conservative in responding to the demand for technical subdivisions as it did to the demand for geographic sections, and it required the same kind of centrifugal pressure to force the Society to act.

In an era characterized by the formation of a great number and variety of specialized societies, it was not possible for ASME to sustain the claims for professional territory made by its founders. Mechanical engineers concerned with heating and ventilating split off in 1894 to establish a national society of their own, and in 1904 engineers interested primarily in refrigeration problems did the same. An even greater shock came the following year when the Society of Automotive Engineers was established, to rival ASME in a new mechanical field of clear future importance. These early defections remained in the corporate memory and were used as a major argument for recognizing the needs of special interest groups. But the Society actually took another fifteen years to work out its attitudes and policies toward technical divisions.

The matter of technical divisions was part of a larger problem. The urge for special interest groups came at the same time as the movement for geographic sections and the ferment in municipal engineering societies for a different style of professional institution. In the interest of maintaining a vigorous organization of their own, ASME's leaders were thus forced to deal simultaneously with three separate but interrelated issues: the structure geographic sections should have, the structure appropriate for technical divisions, and the form of affiliation the Society might develop with strong, existing associations, whether organized for reasons of location or technical specialty. But another factor inhibiting the easy emergence of technical divisions was the range of opinions among the Society's leaders about how much autonomy such groups should have. There were those who favored a set of loose rules, those who would keep a tight rein on the Society's subdivisions, and sometimes, as in the case of the Gas Power Section, there were people on both sides of the fence.

The Gas Power Section grew out of the 1907 annual meeting and a sudden surge of interest in automobiles and the internal combustion engine. Charles E. Lucke, a junior member of the faculty of mechanical engineering at Columbia University, had arranged a special session on gas power for the meeting and it attracted a great deal of notice. But Lucke, Fred Low, H.H. Suplee, R.H. Fernald, and a number of other members had also drafted a petition to the Council calling for the establishment of 'a Professional Section of the Society for the promotion of Gas Power Engineering.'[2] In marked contrast to its previous reluctance to create geographic sections, at its 6

2   ASME Council Minutes 6 Dec. 1907

December 1907 meeting the Council welcomed the petition and immediately appointed a Committee on Affiliated Societies, chaired by Frederick Hutton, to give the request favorable attention. Only a month later, at the Council's next meeting, Hutton's committee reported a set of recommendations for professional sections that emphasized 'broad local self-government.'[3] Thus, under regulations much more favorable than those governing geographic sections, the Gas Power Section was formed.

The time certainly seemed ripe for it. Besides the intrinsic interest of the subject, the Gas Power Section enjoyed some other aids. One was that the Society's leaders had SAE much in mind, and that pushed them to think about the Gas Power Section in terms of regaining lost ground. For instance, in his draft of a new by-law for professional divisions, Jesse Smith used as his example an 'Automobile Section' of ASME. Frederick Hutton, an ardent 'automobilist' himself, entertained in his presidential address the hope that by a kind of affiliation SAE members might be drawn back into the Society's fold. And Calvin Rice reported to Hutton's Committee on Affiliated Societies that some of the industry's most influential men were 'cordial' to the idea of an automobile section of the Society.

The Gas Power Section proved an overnight success and the technical press reacted enthusiastically to it. *Cassier's Magazine* attributed the new section's immediate vitality to the wide interest in the internal combustion engine and published an account of its first meeting. The magazine noted several important features of the event. One was the commercial value of technical specifications in the sale of gas engines, suggesting a need for the same kind of order ASME's standard had provided buyers and sellers of steam-boilers. The national character of the section was emphasized and the magazine contrasted its meeting with those normally held by engineering societies, at which any one paper tended to interest only a few in attendance. Finally, *Cassier's* connected the professional section to a trend in ASME toward simultaneous sessions at its spring and winter meetings. 'The whole movement,' according to the article, 'forms part of the modern system of specialization necessary for the successful accomplishment of large undertakings.'[4]

Hutton's Committee on Affiliated Societies – R.H. Fernald, F.W. Taylor, H.H. Suplee, and A.C. Humphreys – saw the creation of professional sections in much the same light. Their 1908 report, a remarkably prescient and open-minded consideration of the whole issue of ASME's relation to its own

3 Ibid 14 Jan. 1908
4 *Cassier's Magazine* XXXIII (1908) 589–91

subdivisions, as well as its affiliations with already established societies, made several important points. First, it viewed the Gas Power Section as a precedent for others and advised the Council to 'have clearly in mind the advantages to the Society at large which will accrue from the growth and multiplication of specialized sections.'[5] The committee also perceived a crucial difference between geographic and specialized sections. Technical divisions would have their meetings at the usual annual and semi-annual conventions of the Society, and that logically meant simultaneous sessions, papers that met uniform standards of acceptability, co-operation with the Committee on Meetings and Publications, staff support, and Society funding. In that context, section meetings could include non–ASME members without any difficulty and there was no basis for additional section dues.

The central objective of the report was to weave the work of technical sections tightly into the fabric of the Society, rather than to set them apart in separate administrative structures liable to drift into independence. ASME had already lost important segments of its own community, the committee sternly warned, 'because it was not prepared to handle the energies resident in groups desiring specialization.'[6] In contrast, the committee meant its report to inaugurate a new era in Society policy and it therefore also sketched out plans to accomodate associations of mechanical engineering students and of engineers already organized in societies whose objectives ASME shared. The benefits it foresaw in a new spirit of engineering society co-operation are evident in a passage from its report:

Your Committee believes that with the growth of cities, and growth of productive industries in them, the increasing education of those following productive industry and the increasing recognition of productive industry as a factor in the modern community such associations will increase in the future in number and in significance. There are few opportunities within its scope which give larger promise than that of backing up these associations by the power, the prestige and effective co-operation of the American Society, inasmuch as it favors the advance of mechanical engineering at once along educational, social and professional lines.[7]

The immediate and phenomenal success of the Gas Power Section bore out the committee's sense of the pent-up energy within specialized groups. By 1911, besides its executive committee, the section had five different sub-

5 'To the Council' [Draft report of the Committee on Affiliated Societies] n.d.; Taylor Papers
6 Ibid
7 Ibid

committees in action and a nation-wide membership of 365, one-tenth of the Society's total. But despite the broad attractions of the new policy and the Gas Power Section's brilliant success, no other technical sections were formed for the next decade. Indeed, in 1914, the Gas Power Section itself was legislated out of existence by an administrative reorganization the reasons for which are as puzzling as they are obscure.

Some of the Society's leaders clearly favored a more restrictive policy than that of the Committee on Affiliated Societies. For example, Jesse M. Smith, President of the Society in 1908 when its report came out, proposed a set of regulations that would have tied professional sections very closely to headquarters and effectively prohibited non–ASME member participation. And Frederick Hutton, who had raised the concept of affiliated societies in his address as retiring president at the 1907 annual meeting, had actually put forward a much different plan than his committee adopted. As essentially conservative as Smith's by-law, Hutton's recommendations utterly missed the spirit of the movement for geographical and technical sections. The 1905 regulations for local sections, he claimed in a speech on 'The Mechanical Engineer and the Function of the Engineering Society,' had led to meetings that were merely social. He recommended instead an affiliated relationship between local groups and the national society that emphasized the 'prestige of affiliation with the larger body,' and the maintenance of standards in the affiliates 'to a plane of creditable achievement.'[8] ASME would have no financial responsibility for the local groups and their members would not be members of the national society. It was a scheme, in other words, with nothing to recommend it except Hutton's sense of *noblesse oblige*.

That kind of thinking suggests the Gas Power Section was an anomaly, and subsequent events support this proposition. In 1909, fresh in the wake of its early success, another petition was presented to the Council calling for the formation of a machine shop section. At its meeting of 13 April 1909 the Council referred the matter to a special committee, consisting of the former members of the Committee on Affiliated Societies, for their consideration. Frederick Hutton also served as chairman of this committee and at the Council's next meeting he reported that his committee was making progress. But between that point and the November 1909 meeting of the Council, the petition was shifted to the jurisdiction of the Executive Committee of Council, which reported its conclusions.

8 Frederick Remsen Hutton *A History of the American Society of Mechanical Engineers from 1880 to 1915* (New York: ASME 1915) 341

The Executive Committee first itemized the advantages of professional sections. They captured new ideas and led to the more active presentation of papers. Furthermore, the concentration of experts in a given field enhanced debate. Dividing papers into topical groups also spared members the inconvenience of listening to papers outside their interests. Finally, sections would tend to attract people who would not normally contribute to meetings and in the same way they would draw into Society membership men who would not otherwise have joined.

Against these advantages, briefly enumerated, the committee put forward a number of disadvantages. In an argument much like Frederick Hutton's claim that geographic sections would be reduced to social gatherings as soon as local talent for papers was exhausted, the committee declared that once everyone knew the most recent advances in a field, it would be hard to maintain enthusiasm. In that case, the committee asserted, 'the policy of the general meeting with its broader and more catholic interests becomes the more permanent and easily attainable one.'[9] The committee also stated that a specialist actually performed better before a general audience, on the grounds that a person would try harder and be forced to focus his remarks more sharply. Besides, one could achieve the effect of a section by simultaneous sessions. Furthermore, the committee expressed its concern over non–ASME members of the section; 'It is not yet proved that the predominance of such affiliated members in any section is more of an advantage than a possible detriment if they happen to be out of harmony with the ideals of the Society.'[10] Another disadvantage, in the committee's judgment, was that papers offered only in section meetings would be denied the review of those with related experience.

On the strength of its analysis of the pros and cons, the Executive Committee recommended against the formation of a machine shop section, suggesting instead that a subcommittee of the meetings committee be created to help plan sessions at the annual and semi-annual gatherings of the Society for those members especially interested in that topic. In complete reversal of the stance taken by the Committee on Affiliated Societies, these recommendations became general Society policy. In other words, at almost the same time that geographic sections were winning their independence from the meetings committee, the Council decided that all technical interest groups would be formed as subcommittees of the meetings committee, rather than as professional sections. And that policy remained in place for the next ten

9 Council Minutes 9 Nov. 1909
10 Ibid

years, until 1919. In February of that year, without a formal vote, the Council reaffirmed its 'present standing policy' that instead of professional sections, special committees would be appointed 'and thus simplify the Society's organization.'[11]

Of all the factors that influenced the Council's judgment in the matter of professional sections, it is most easy to recognize its fear of non–ASME members and an unflattering concern for the ability of professional sections to sustain themselves intellectually. But beyond the Council's conservatism and its suspicions, there were difficult balances to be struck in the administration of technical interest groups. William Kent and others long familiar with the Society had already pointed out, for example, that many technical papers were scarcely disguised advertisements, and it was easy to imagine that commercial interest might be even more active within sessions devoted to a particular industry. Thus the Society had to find workable procedures that stimulated papers for technical sessions yet allowed for control of their content. ASME's leaders had also to face realistically the fact, which the Committee on Affiliated Societies recognized, that to many specialists there were attractions and advantages in societies of their own. This was another situation calling for sensitive procedures to balance the Society's interests and those of technical specialists. And among those less obvious influences shaping Council action, some members felt that, while specialization was intimately associated with technical progress, an important Society objective was to provide linkages between bodies of information and bring about a kind of intellectual synergism.

Although ASME's new policy on professional groupings stood out in sharp contrast to its immediate predecessor, it did not seem an entirely retrograde step to the technical community. Subcommittees were rather quickly formed in the fields of management, machine shop practice, textiles, and cement, and their creation brought a favorable response in the technical press. The *American Machinist*, which had sharply criticized the Society for its handling of the Milwaukee episode, described the new policy for technical interest groups in an article entitled 'Forward Movement of Mechanical Engineers.' The magazine claimed that ASME's scope was already wider than that of any other national society and that the purpose of the new policy was to extend its usefulness and influence still further. Special subcommittees would investigate the problems of varied fields and study their new developments 'with the purpose of keeping the members fully informed as to the best of recent technical developments, and permit

11 Ibid 22 Feb. 1919

of applying solutions of problems in one field to similar problems in other fields.'[12]

Over the next few years, a number of additional special interest groups were formed. In this fashion, in 1912, the Council added subcommittees on air machinery, fire protection, hoisting and conveying, industrial buildings, iron and steel, and railroads. The following year another committee was formed, on depreciation and obsolescence, and the Gas Power Section was converted to a subcommittee; and in 1915, a subcommittee on the protection of industrial workers was created. But these groups more directly suited the subject categories of meeting planners than they reflected a logical division of technical interests. Several of the subcommittees lasted only a short time. The number and names changed, as if the dominant consideration was again the organization of sessions for the Society's meetings rather than the creation of a permanent structure. These measures only momentarily satisfied members seeking some accomodation for special fields of interests.

In the immediate post-war period a new reform thrust recaptured the kind of enthusiasm and energy the Gas Power Section had so successfully stimulated in 1908. Morris L. Cooke's efforts to democratize the Society and broaden its sense of social responsibility were undoubtedly part of the new spirit, but it is also clear that the First World War had a substantial impact on the organization. The popular expression that it had been an 'engineer's war' gave the profession a sense of central importance in that heroic drama, while the government's mobilization and management of the nation's transportation system and its industrial resources suggested that the engineering approach might find a whole new set of peace-time applications. That feeling was reflected in a declaration of principles adopted at a business meeting of the Society in 1919. As one resolution proclaimed, 'Credit capital represents the productive ability of the community and should be administered with the sole view to the economy of productive power.' The engineer's role in this new planned state was also indicated in another resolution, 'Every important enterprise must adopt competent productive management, unbiased by special privilege of capital or of labor.'[13]

It was somewhat in that spirit that the Committee on Aims and Organization was created in 1918. Its charge was to formulate ASME's future objectives and programs 'in the light of modern development and present day

12 *American Machinist* XXXV 2 (1911) 1043
13 Council Minutes 5 Dec. 1919

thought.'[14] The committee took that broad mandate seriously and in its investigations over the next several months considered such diverse issues as the Society's role in solving industrial unrest and its participation in a national organization of engineers to bring the force of the entire profession to bear on crucial national issues. In an article he wrote for *Mechanical Engineering* in 1919, the committee's chairman, Louis C. Marburg, described the intellectual context of its work. The war had brought revolutionary changes, he said, that challenged traditional notions of individualism, business, and the role of the state. In that critical atmosphere, more and more people were doubting long-held theories and casting off accustomed methods. As a result, Marburg claimed, 'a momentum has been acquired by this habit of thought which will carry us, if we but guide it wisely, far into the promised land of economic efficiency and social justice.' Translating that sense of things into an administrative structure for ASME, he suggested 'that possibly two entirely different types of organization may be wanted, one for professional work and one for public activities.'[15]

Marburg's committee devoted most of its attention to the establishment of an engineering association of the second type that would deal with the profession's public responsibilities, an effort which led to the creation of the Federation of American Engineering Societies. But among his committee's recommendations was one calling for the appointment of a Committee on Professional Sections, 'analogous to that on Local Sections,' and at the annual meeting in 1919, the Council authorized the formation of technical sections.[16] Charles E. Lucke had been working the same direction from his position within the Committee on Meetings, after the Gas Power Section was absorbed into it. A society conducting its affairs as if everything appealed to everyone, he argued, 'can never attain a degree of usefulness commensurate with the interests represented by its members.'[17] According to Lucke, technical divisions were the natural result of the diverse activities of mechanical engineers and he warned that, if professional sections were not organized, members would split off in groups of their own, as refrigerating and automotive engineers had done. Lucke made another familiar claim when he argued that a national organization made possible a degree of intellectual cross-fertilization smaller societies could not achieve, and he suggested that some of them might be better off as ASME technical sections.

14 Ibid 20 Sept. 1918
15 *Mechanical Engineering* (Jan. 1919) 14
16 Council Minutes 24 Oct. 1919
17 *Mechanical Engineering* (1920) 303–4

Through these old arguments and Marburg's appeal for change, ASME was finally brought back to the idea of relatively independent technical divisions. The committee that worked out the details for their administration reported its conclusions at the first Council meeting of 1920. Besides setting out the internal organization of sections, the committee's most important recommendations were that non–ASME members could join sections as affiliates and that sections had a legitimate call upon the Society's treasury. These and other details, some of which recapitulated the structure of the Gas Power Section, gave the Society's technical divisions a position that was at least constitutionally equivalent to that of its geographic sections and, in the creation of a parallel Standing Committee on Professional Sections, the technical interest groups also gained the right to sit in at Council meetings.

The technical sections created in 1920 – aeronautics, cement, fuels, gas power, industrial engineering, machine shop, ordnance, power, railroads, and textiles – were seen by the Committee on Professional Sections as ASME's proper territory. The committee put that concept forward at several points in its report to Council in January 1920. Edwin B. Katte, its chairman, talked of mechanical engineering fields 'fast being invaded' by other organizations and of 'the most pressing need' for an ordnance section 'to forestall the formation of an Army Ordnance Society.' Katte concluded his report on the same theme, arguing for the prompt establishment of professional sections both 'to encourage greater technical activity on the part of the specialized members of the Society and to forestall the invasion by less qualified organizations of the established and well recognized provinces' of ASME.[18] That notion was useful as a defensive strategy against the raids of other associations, but it left unresolved the jurisdictional lines between sections, and that proved a problem. There were several name changes, as divisions looked for the right boundaries for their own concerns, and several divisions – cement, ammunition, national defense, and rubber and plastic – proved to be short-lived. Others rose up during the 1920s in their place: forest products, printing machinery, oil and gas power, petroleum, hydraulics, iron and steel, and applied mechanics.

Besides the problem of unclear assignment of responsibility, there were significant differences between the programs of divisions, and by 1923 it seemed that some umbrella structure might be necessary. One suggestion was that there could be a few large divisions, (like the Society's later departments), plus a number of smaller and perhaps temporary professional groups – a plan that also resembled the previous organization of subcommittees under the

18 Council Minutes 24 Jan. 1920

Committee on Meetings. The same general concept surfaced again in 1924: the technical sections would be grouped into five large units – administrative, power, metal industries, non-metal industries, and transportation. And in 1925 the Committee on Professional Sections reviewed the entire program, seeking to define clear objectives and a rational administration.

These organizational struggles did not arise from any particular failure of the Society's technical purpose – indeed the 1920s were a period of great vitality in that sphere of action. A special committee formed just after the war co-ordinated an expanding array of research projects and within the next decade had raised and spent $200,000 to conduct them. ASME's boiler code, begun in 1914, had by 1928 grown to eight sections; there were twenty different power test codes committees, an equal number of groups dealing with various aspects of safety, and more than 200 committees working in the area of standards. So there was no lack of enthusiasm or of willing hands. The problem was rather that these activities still did not resolve some inherent difficulties that had been apparent from the first discussion of technical sections.

The 1925 review, for example, considered the administrative structure of professional divisions, their relation to headquarters, and their connections to local sections. In its analysis of their functions, the Committee on Professional Divisions pointed out to the Council four major contributions of technical interest groups – they generated papers for the Society's annual and semi-annual meetings, they developed technical programs for local sections, they produced specialized articles for publication, and they helped the Society co-operate with other groups and industries having related interests. The Oil and Gas Power Division, for instance, had organized a week-long conference, largely funded by industry, that proved a great success. Sixteen organizations, including the American Chemical Society, the United States Geological Survey, the American Petroleum Institute, and the National Bureau of Standards, participated in the planning for 'Oil and Gas Power Week,' making it a model of institutional co-operation. 'Management Week' was another example of the kind of co-operation technical divisions could achieve. And these efforts of the Society's professional divisions, the committee claimed, were matched by their attempts to co-operate with local sections. For instance, since it proved difficult for each of the Society's sections to appoint representatives to every technical division, the Committee on Technical Divisions had created a few large groupings according to major subject fields in order to make it easier for sections to get help in planning their programs. In a parallel fashion, the Annual Progress Conference, begun in 1924, served 'to bring together periodically all the agencies in the Society having to do with the preparation of the Society's technical program.'[19]

All this labor, the committee reminded Council, cost ASME practically nothing. The time members gave to technical division work was either their own or the contribution of their employers, meetings and papers were already paid for, and even the administrative chores were handled by a system of volunteer secretaries worked out by W.E. Bullock of the headquarters staff, making it possible for one assistant secretary to supervise the entire operation of technical sections. It was clear from the committee's report that the Society got its technical divisions at a bargain, but it soon became equally apparent that things would have to change. Professional sections had grown beyond the support system they started with and their ability to defend the Society's territory against invading organizations appeared to be seriously threatened.

Those who addressed themselves to the problems of the professional divisions often sketched a relation to local sections that implied complementarity and symmetry. Local sections gave grass-roots solidity, technical divisions provided specialized knowledge. Each made an equally important place for individuals in an otherwise far-flung and diverse organization. While no one ever thought the number of technical interest groups would come to equal that of geographic sections, practically all organizational discussions of technical subdivisions described a structure parallel to the one for geographic units. In practice, however, an equivalence between those two major arenas of ASME activity required not only matching constitutional provisions, but equality in funding, staff support, and rewards for service.

These other evidences of a lack of parity for the Society's technical sections became particularly important in the war for territory. W.A. Hanley, a former chairman of the Committee on Local Sections and later president of the Society, wrote to Pierce Wetter, the headquarters staff man responsible for technical divisions, about the whole problem as it appeared to him in 1929. The strategy of forming special interest groups to head off other competing organizations had been principally responsible for the present technical division structure, Hanley said. The ASME could never hope to stop the formation of every group in its broad field of interests since many of them, even with the word engineering in their title, were actually trade associations concerned with both technical and commercial issues. But in the general realm of engineering, and assuming the country to be 'better served by a large society in contrast to many small societies,' Hanley claimed the techni-

19 Ibid 16 March 1925

cal divisions could adequately protect the Society's interests only if they had more power within the organization. He drew a parallel with the local sections movement that had given 'a geographic spread to the society and a prestige, which it did not formerly have.' Hanley saw the same potential in an expanded role for technical divisions. By increasing the number of industries represented ASME would also increase its activities and membership; 'we are the society of the industries,' he reminded Wetter, 'and the industries are numerous.'[20]

Ralph Flanders, then a member of Council, took particular interest in the problem. 'I suppose in a sense it is true that this project is "my baby,"' he wrote to Colonel Paul Doty, chairman of the Committee on Local Sections in 1929, and subsequently Society president.[21] Flanders claimed the spreading profusion of mechanical engineering societies was a serious matter, but his letter also revealed that the efforts to resolve it had raised some delicate issues. One proposal, for instance, was to give more prestige and standing to the work of professional divisions by elevating their chairmen to the rank of vice president and thus increasing their power on the Council. That idea had stirred up a great deal of interest, tending both to displace the more general issue and to polarize opinion. Indeed Doty, who called it 'the Vice-Presidential grab,' set himself strongly against the notion. It smacked of 'special privilege' and of 'class distinction,' he claimed, and was unconstitutional and unnecessary as well.[22] Instead, Doty suggested increasing the number of seats on Council and making three vice-presidential nominations from names presented by the professional divisions. Since the nominations process was pretty well in the hands of the Committee on Local Sections, Doty's scheme was at best a limited step towards parity for professional divisions.

In a personal reply to Doty, Flanders attempted to defuse the whole affair. 'These are simply two thoroughly useful and effective cross sections of the Society on different planes,' he said, and claimed that his concern was not so much the creation of professional division vice presidencies, as it was to combine the country's mechanical engineering activity 'into as compact a body as possible, preferably under the banner of the ASME.'[23] As he saw it, the problem was to discover the best way to extend the power and scope of professional divisions with the fewest changes in procedure or constitution, in order to create 'the maximum of advantages in bidding for the support

20 W.A. Hanley to P.T. Wetter, 20 March 1929; Flanders Papers, Syracuse University, Syracuse, NY
21 Ralph E. Flanders to Col. Paul Doty, 30 April 1929; Flanders Papers
22 Col. Paul Doty to Ralph E. Flanders, 18 April 1929; Flanders Papers
23 Ralph E. Flanders to Col. Paul Doty, 30 April 1929; Flanders Papers

now going or likely to go toward independent organizations.' Flanders wished also to discover in a quiet way which societies already in existence might be amalgamated into ASME, and he suggested the technical divisions themselves might be the best group to find the answers to those questions. In any event, he concluded, while the matter naturally raised 'some thought of the proper balance' between professional divisions and local sections, more was at stake than a quarrel over 'internal policy.'[24]

In fact, of course, the affair had a great deal to do with internal policy. The Society had enjoyed a very substantial increase in membership with the growth of geographic sections, while by 1929 technical divisions still had not reached the same level of 'take-off,' despite a great deal of enthusiasm for them in 1919. Victor Azbe, chairman of the Fuels Division, explained what he thought the problems were in a letter Ralph Flanders described as a 'dandy.' Azbe admitted his own division was 'a rather lame' one. It had good men in its top posts, but somehow it lacked 'spirit.' However, he saw his division's difficulties as symptomatic of some larger problems common to all the technical groups. In the first place, he claimed, a man busy and important enough to serve on a divisional executive committee tended to see it as just another committee assignment. A vice presidency for the chairman might help the situation a bit, but even more important, Azbe said – putting his finger on a fundamental truth for volunteer organizations – was 'to play up the Executive Officers, tickle their vanity, make it worth their while to work for the Division and make others anxious to work themselves up in the Division.'[25] Azbe argued that the headquarters organization was at fault, too. Its publicity was 'rotten,' and no one on staff knew anything about the industry with which his division was concerned. But in his catalog of complaints, Azbe struck a new note, pointing out the crucial role of staff. For instance, he said that if Pierce Wetter could give a significant amount of his time to aeronautics, instead of having to care for all the divisions, the formation of a separate society in that field might just be prevented. What ASME needed was 'adequate staff, adequately paid, with definite duties.' When the men from local sections complained, it was about the amount of money they were allocated. But in 1929, members of the technical divisions had an even more basic set of needs to be met.

The creation of a special Revenue Committee in 1927 gave some hope to those who wished to see technical divisions occupy a more important place in the Society's affairs. A sudden decrease in membership in the last half of the

24 Ibid
25 Victor J. Azbe to C.E. Davies, 22 Aug. 1929; Flanders Papers

1920s had raised the question, 'has the Society reached the end of its growth?' and the committee was formed to study that possibility. Chaired by William L. Batt, it included Dexter S. Kimball, Ralph Flanders, Conrad Lauer, Erik Oberg, James D. Cunningham, and J.L. Walsh. Five of those seven became ASME presidents and, in an association which tended to identify its chief officers well in advance, the committee's composition indicates the importance attached to it. Batt argued that in contrast to the organization's normal expansion through geographic sections, there were 'unlimited opportunities for the development of the Society in professional lines,' and he stressed the need for urgent attention to that activity as a solution to the membership problem.[26] It was also from the Revenue Committee that Batt and Flanders floated their idea of technical division vice presidencies, which they made part of a more general scheme to refashion the Council into a deliberative rather than an administrative body, shifting more of those latter concerns to the Executive Committee of Council. And one of the important elements in this plan was 'a more scientific preparation of the budget.' When faced with the real facts of the Society's future possibilities for income, Batt claimed, the Council 'cannot longer continue to give Local Sections everything they ask for and content Professional Divisions with what few dollars happen to remain in the treasury.'[27]

Despite the eagerness of men such as Batt and Flanders to even the political balance between technical divisions and geographic sections, Council's rejection of their proposal to create technical division vice presidencies and Doty's reaction to the idea suggest that local sections maintained an important edge. So one of the handicaps technical sections faced was that they were unable to compete as successfully as local sections for a larger allocation of the Society's resources.

Another difficulty was that within the sphere of technical divisions a similar kind of political inequality prevailed. In his 1929 report to Council, Archibald Black, chairman of the Commitee on Professional Divisions, argued that the demand for specialized technical groups had presented a historically important choice. The organization had to decide whether to be 'a society of steam engineers or to cover the whole mechanical field.'[28] Others saw technical division politics in much the same light. E.P. Hulse, chairman of the Printing Industries Division, one of the smallest in the Society, argued that ASME's natural '"bent" toward steam boilers and valves as a chief field of

26 Council Minutes 29 March 1929
27 W.L. Batt to Ralph E. Flanders, 18 May 1929; Flanders Papers
28 Council Minutes 2 Dec. 1929

mechanical engineering' had made it difficult for divisions such as his to get started.[29] W.C. Glass remembered the same thing: 'I had been a member of the Society for years and got nothing out of it except "steam." I had no interest in steam, as such, and after Council reluctantly gave in to a few of us who wanted recognition for our industry we went ahead more or less successfully.'[30]

From the beginning management, power, and machine shop practice had been the dominant technical divisions and they greatly exceeded most of the others in membership. In 1921, for example, the Power Divison enrolled three times as many members as the Railroad Division and four times the number in textiles or aeronautics. Moreover, these three largest divisions were, comparatively, even bigger in 1936, by which time the Fuels and Steam Power Division had grown to almost ten times the size of the Railroad Division and had twenty times as many members as textiles. In sheer numbers, if nothing else, certain fields of mechanical engineering dominated the others and the weight of their influence in ASME's sphere of technical purpose was a bit like that of New York in the realm of geography. And it was not easy to break through the control of these central interests, as railroad men in the Society had long before learned. But the Petroleum Divison, boisterous and geographically concentrated, provides a good example of how, with the extra benefit of timing, a cohesive group could force its interests on Council.

Just as in the mining industry forty years earlier, American oilmen in the 1920s discovered the relation between technical knowledge and increased production. And, in just as dramatic a fashion, crude petroleum output in the United States reflected the connection, more than doubling in the decade from 1920 to 1930. Thus, in 1930, when W.G. Skelly, president of Skelly Oil Company, contrasted pioneer techniques in his industry with the mechanical engineering skills necessary for contemporary practices, it appeared to him that the changes had come hard and fast. Less than ten years ago, he said 'we just conducted the oil business kind of as a rule of thumb.'[31] But more had been accomplished in the past five years than ever before through the application of engineering knowledge. The time to drill a well had been reduced

29 E.P. Hulse to Archibald Black, 26 Nov. 1929; Flanders Papers
30 W.C. Glass to P.T. Wetter, n.d.; G.B. Pegram Papers, Columbia University, New York, NY
31 Summary of the Four Day Meeting of Council Committee on Local Sections, Mid-Continent Section, ASME and Petroleum Division, ASME and Executives of the Petroleum Industry Held at Tulsa, March 11, 1930; Flanders Papers

by one-third while the amount of equipment had tripled. The need for specialization and technical education had brought a different sort of person into the oil business, Skelly claimed, and he predicted its future 'will be largely worked out through engineers.' But if those rapid changes seemed surprising to Skelly, they caught ASME flat-footed. Focused more on steam-power than petroleum and remote from the most dramatic western developments in the industry, the Society's leaders were unaware of its technical requirements or of its needs for institutional support.

The Petroleum Division was formed in 1924, on the initiative of members of the Mid-Continent Section in Tulsa. From the outset, it enjoyed considerable autonomy, probably because its concerns and locations were so far from the active business of the Society's New York headquarters. But as the Batt-Flanders crusade to gain more recognition for technical divisions gathered momentum in 1929, Tulsa members began to push for a greater sensitivity to their interests. The secretary of the Tulsa section sent the Council a telegram in November 1929 asking for a meeting 'to obtain adequate activity by the Society on matters pertinent to the promotion of petroleum progress from the mechanical engineer's viewpoint.'[32] But that was only the first step. Besides the combined pressure of the Petroleum Division and the Mid-Continent Section, Tulsa mechanical engineers used the threat of affiliation with competing societies to force the Council to action. Ernest Hartford, secretary of the Committee on Local Sections, spent several weeks there early in 1930 and he wrote back to New York warning of the danger AIME posed to the Society's interests and of the need for a better awareness of the amount of mechanical equipment the petroleum industry used. Like a battlefield scout, Hartford assessed the enemy's strength and movements for headquarters. 'While we have been dormant,' he wrote Calvin Rice, 'the Amer. Inst. Mining Engineers have been real active here.' In Hartford's judgment, the American Petroleum Institute did not pose such a threat but, as he told the Committee on Local Sections, the mining engineers 'have us absolutely stopped at this time.'[33]

By February the problem had become 'an acute situation' and the Council agreed to a major meeting in Tulsa in an attempt to resolve it. Their own discussions of the matter reveal some of its dimensions. One source of discontent was that ASME publications carried few articles touching the interests and problems of the industry. That was a general feature of technical division

32 Council Minutes 14 Oct. 1930
33 Ernest Hartford to Calvin W. Rice, 6 Feb. 1930; and Ernest Hartford to the Committee on Local Sections, 6 Feb. 1930; Flanders Papers

politics: strong divisions commanded more attention at meetings and space in *Mechanical Engineering* and the *Transactions*. But Calvin Rice mentioned in particular the oilmen's complaints about the ASME boiler code. Petroleum interests resented the almost exclusive attention to steam-power, to the neglect of the development of specifications for unfired pressure vessels. Further, the oilmen had a totally different approach to safety factors and to the amortization of capital equipment from that of the steam-boiler industry. Either of those issues might have been touched in the 'fundamental modifications in the Code' which Rice advised the Council had been made 'to take care of the oil industry.'[34]

Eugene W. O'Brien, editor of *Southern Power Journal*, had occasion to travel through the Oklahoma oil territory as the region's representative on the 1930 Nominating Committee and he sent Calvin Rice his own candid assessment of things. The problem was not a new one, he reminded Rice, and not unknown to Council, 'but it seems particularly acute at this time.' The trouble, O'Brien said, lay in the Society's failure to provide technical support to the petroleum industry. Oilmen continually complained of the lack of papers or activities of interest to them, and charged that whatever was done came too late. 'Among specific claims,' O'Brien said

were that the Society had not kept up with and aided the industry in setting up standards, it had not joined in the study of the many problems of mechanics that beset the new industry, such as those of accurate drilling, etc.; that it had fallen hopelessly behind in the development of the art of building and welding pressure equipment (piping and containers for 2200 lb. gas pressure are now built by welding, which the Society has not yet approved for much lower pressure;) etc.[35]

While the tone of O'Brien's letter suggested a situation that might already have gone too far, the oilmen in Tulsa certainly saw the upcoming meeting as a last chance. Arthur J. Kerr, chairman of the Mid-Continent Section, sent Ralph Flanders a telegram which clearly carried that message: 'Continuance of Society activity in Mid-Continent Section dependent upon Council hearing our case and taking favorable action upon it. Therefore, we want a quorum of council and your attendance in Tulsa next week is believed essential.'[36]

34 [Calvin W. Rice] Secretary to the Members of the Council, 11 Feb. 1930; Flanders Papers
35 W.E. O'Brien to Calvin W. Rice, 3 March 1930; Flanders Papers
36 Arthur J. Kerr to Ralph Flanders, 4 March 1930; Flanders Papers

The Petroleum Division presumably represented Pennsylvania and California oil interests, as well as those from Oklahoma and elsewhere, but it was the Mid-Continent Section that brought about the remarkable pilgrimmage of Calvin Rice, a quorum of Council, and the entire Committee on Local Sections. Just as the Society was beginning to understand the implications of the Great Depression, and with its fiftieth anniversary celebrations only a month away, many of ASME's leading figures entrained for Tulsa to learn the demands of the oilmen. Ernest Hartford had carefully arranged a four-day program with the local section that was designed to educate Council on the equipment of the petroleum industry and the scale of its operations, the codes and standards work it needed, and the kind of staff support the Society might provide. The meeting included oilfield inspection tours, luncheons with Tulsa's financial community, conferences with leading executives of the oil companies, and social receptions in the evenings. At the end of it a staff member was stationed in Tulsa to provide more effective liaison with headquarters, while the Council referred the problem to a joint committee of the Committee on Local Sections and the Committee on Professional Divisions. In June 1930, the Council adopted its report and set up an advisory committee, voted funds to cover the Society's petroleum activities, and decided that 'a working majority of the Petroleum Division Executive Committee shall for the present be from the Mid-Continent Section.'[37]

From one point of view, the 'Tulsa Affair' was a triumph for Ernest Hartford, for those in ASME eager for a more aggresive policy of technical division development, and for the members of the Mid-Continent Section who by their own forceful energy had captured the Petroleum Division. And there were some other beneficiaries, even if unwillingly so. The boiler code committee, long dominated by a small group of steam-boiler men, was pushed into dealing with the growing importance of unfired pressure vessels and welding technology. Yet the Council's willingness to commit leadership, staff, and funding to the Petroleum Division did not solve all of the Tulsa group's problems or those of technical divisions generally. Indeed, some of the issues that had proved troublesome at the beginning would persist as essential elements in the nature of technical division activity.

Just as with ASME's social economy, its technical purpose was characterized by a continuing need to strike balances, to discover workable procedures for a set of functions constantly subjected to contradictory pressures and defying any final solution. And throughout the depression years of the 1930s these

37 Council Minutes 9 June 1930

difficulties technical divisions experienced were exacerbated by financial stringency and a relative lack of political power within ASME. Successive chairmen of the Committee on Professional Divisions spoke, for instance, of the responsibility to provide papers for the Society's meetings and publications. But it proved a constant struggle to weed out papers that were too commercial and there was now less money to print even the good ones. In the interest of ASME's prestige, W.A. Shoudy argued for a 'hard-boiled' policy of rejecting anything except 'thoroughly fundamental papers.'[38] However, that was no easy solution, since technical divisions measured the number of papers in their field as evidence of headquarters support.

In 1935 W.H. Carrier, of air-conditioning fame and a man of strong opinions, wrote to his fellow members of the Committee on Professional Divisions a long and thoughtful letter which put some of these difficult issues in plain language. He began with the most general considerations. Was ASME to be a sort of 'common meeting ground' for mechanical engineers of widely diverse interests, or was it rather to be a federation of 'self-contained' and specialized technical divisions? The technical advances of the previous half-century had brought a great deal of specialization and many specialized societies, Carrier pointed out. However, he said, the question was whether that kind of specialization was desirable and what ASME's policy about it should be: 'Shall we divide our whole society up into groups of this character, or shall we look to this Society to provide something for the engineers regardless of their affiliations which cannot be found elsewhere?'[39] The way Carrier framed the alternatives telegraphed his own choice. The latter approach, he said, was 'the only logical one, and the only wise one.'

Carrier's plan reflected the deep longing of many engineers for a professional society elevated above commercial interests and devoted only to some transcendent increase and diffusion of technical knowledge. It was as if he had imagined the Holley-Thurston vision, but without the business connection. Actually, there was a greater similarity between Carrier's ideas and the society Frederick Hutton had described in his 1907 presidential address. According to Carrier, ASME should be 'a society above societies,' or as he subsequently put it, 'an Engineering Academy.' He saw the formation of specialized groups as natural and desirable, but their inevitable tendency to become trade associations made them suitable only for something like affiliate status in a relation that should cost the Society nothing.

38 Minutes of the Standing Committee on Professional Divisions 25 April 1936
39 Ibid 4 Oct. 1936

In Carrier's opinion, most of the Society's technical divisions were organized on an incorrect principle and had also proved faulty in practice. They were narrow, clannish, ineffective, expensive, and a detriment to the Society's prestige, he claimed in a remarkable twelve-point bill of particulars. He criticized the limitation of knowledge to narrow channels, the poor quality of papers generated by the system, and a related tendency toward technical advertising. As a consequence, Carrier pithily concluded, the Society had been reduced to 'peddling peanuts.' In contrast, he proposed a different theory of organization for professional divisions, according to the fundamental subject categories of applied physics. By that Carrier meant such 'natural' units as mechanics of fluids, thermal engineering, or combustion engineering, which would identify 'horizontal' professional divisions within the Society. Thus, members of the mechanics of fluids division might variously be interested in aeronautics, pumps, or compressors, but they would share a common basis for their interests and could thus communicate in mutually understandable language. This 'cross-fertilization' of ideas not only fulfilled ASME's most useful purpose, Carrier asserted, but it was the best stimulus to technical progress, and he brought the history of technology to his aid to prove how often technical breakthroughs in one area were the result of observations drawn from a related field.

Even if in an exaggerated way, Carrier's criticisms of the technical divisions highlighted the worst of their problems. And some of his ideas found indirect expression in a 1936 reorganization of the divisions into five departments – basic science, power, transportation, manufacturing, and management. But the professional divisions were the outgrowth of a set of inherent circumstances that remained unchanged, thus rendering other of Carrier's suggestions unworkable. For example, the centrifugal pressure created by segments of the membership ready to fly off into independence if their own interests were not recognized remained a real and continuing problem. Just as Carrier was presenting his own theories, for instance, the Oil and Gas Power Division threatened to form a separate Diesel Users' Association outside the Society if Council did not meet their demands for increased funding and for control over the terms by which non–ASME members were to be associated with the division.

At the same time Pierce Wetter was warning the chairman of the Committee on Professional Divisions that the members of the Graphic Arts Division were equally restive and pointing out that the Institute of Aeronautical Sciences was already lost to the Society's sphere of influence. Wetter, whose long struggle for the well-being of the technical divisions had given him some hard-won experience, could see the impact of the Society's financial policies. 'Someone had said,' he reported to W.A. Shoudy, 'that we treat our divisions as if they were our children, and not yet sufficiently grown up to be trusted.'[40] The Society's way of doling out money to cover a division's expenses was bad enough, Wetter said, but worse was the headquarters practice of keeping the surplus from any sales of division-generated pamphlets or papers. Furthermore, budget cut-backs had reduced the annual allotment per section to $150 and he claimed that was simply too little for effective work. The fact is that local sections were also hard-pressed during the depression years and their budgets were cut, too. But they had an important advantage; ASME's leaders, as one put it, thought of the geographic sections as having 'a first mortgage' on the organization's income.

At the end of 1936 Pierce Wetter resigned, after having finished ten years as the staff member responsible for technical divisions. Assistant Secretary Clarence Davies had predicted 'a spendid future' for Wetter when he filled out his staff rating form in 1929, but a harsh quarrel with Lewis Sillcox of the Railroad Division and other strains of the job led Wetter to decide he had had enough. Depression retrenchments had reduced his salary by one-third and because he had also lost two-thirds of his own office staff, his work-load had been substantially increased. But he probably knew the circumstances of the technical divisions better than anyone and before he left the society to find another career, he sent Kenneth Condit, chairman of the Committee on Professional Divisions, his own analysis of the situation.

The original concept for technical divisions, Wetter said, was that of small, independent societies within ASME. They were to have staff support and financial aid for separate national meetings, as well as the authority to organize sessions at the Society's annual and semi-annual meetings. Divisions were also to have an important role to play in identifying and organizing fruitful research programs and, in the same vein, they were to point out useful standardization projects. These two roles, plus the annual review of the field, were designed to keep the Society abreast of specialized technical developments and thus make it more able to respond to the needs of specialized groups. Finally, Wetter noted that the professional divisions were to assist local sections with their programs, especially by helping them find speakers on special topics. These extra activities cost the Society money of course, both for staff and for considerably higher printing bills, but the program had been a successful one, he claimed, with an admirably co-ordinated publication scheme.

The depression had wrecked all of that. According to Wetter, it had been possible to create some makeshifts, but the attrition of staff made it practically

40   P.T. Wetter to W.A. Shoudy, 28 June 1935; Flanders Papers

impossible to continue under the original terms. The question, therefore, was what arrangements should divisions have in order to survive existing conditions? In Wetter's mind, the only solution was to let divisions charge dues and keep the revenue. But in order not to make that an extra cost to members, he proposed the associate member rank be used for those primarily interested in the work of technical divisions. For one-half the dues of a full membership, the associate would receive a limited set of Society benefits and the advantages of whatever divisions he joined on payment of their dues. Anyone who wanted the status of member would receive all its privileges and the right to register in three divisions, with a portion of his dues going to them.

Wetter claimed his plan was not only practical, but that it met the objections to technical divisions that were still widespread within the ASME. He identified two principal opposition camps. One group 'wanted to see the Society return to its early days and be a select exclusive group, holding at its annual meeting only one session at a time, containing very distinguished papers.'[41] Those sentiments were still much alive, Wetter said, thinking perhaps of W.H. Carrier, but he argued that distinguished papers came out of specialization and that, without the divisions, papers before the Society would have increasingly to be from non-members.

The other opposition group Wetter likened to parents who feared giving their children a healthy independence. Council parcelled out an allowance from the divisions' own wages, but then took back the unspent portion, and they could never develop long-term programs or the responsibility to carry them out. The divisions might have gotten better support 'had they been organized politically to watch their interests and make demands,' but Wetter claimed they were not interested in 'playing politics.'[42] They were, of course, but they just were not as successful at it. In contrast to the political strength of the men in the Society's geographical sections, only two of the top leaders in the technical units in 1940, for instance, made it to the Council in their subsequent careers. But it was also clear that ASME's internal political struggles were not simply between the two great fields of action. In the technical sphere the large special interest groups enjoyed several advantages over the smaller ones. Management, power, and machine shop practice, as they were originally called, not only enrolled more members, but were also closer to headquarters and enjoyed traditional standing.

From the Council's point of view, and particularly at a time when staff reductions threatened all phases of the Society's operations, it made some

41 P.T. Wetter to K.H. Condit, 16 Nov. 1938; Flanders Papers
42 Ibid

sense to force the divisions to solve their own problems, including their financial ones. Under those circumstances, it became a matter of finding the right balance, of providing just enough support to keep the divisions afloat. But it was a sensitive point, from practically any angle. Council could easily assume, for instance, that divisions had a better chance of industrial support than local sections and allocate the budget accordingly. However, the Society was also concerned to control the commercial impulses of the divisions, and the exhibitions and other activities which raised funds for them raised that troublesome issue, too. At an even more basic level, the Society depended on membership dues for its income and member enthusiasm for its vitality; further, the sharp decline in dues receipts during the '30s made it even more important to keep members instead of driving them away.

All these issues are reflected in the records of the Committee on Professional Divisions throughout the remainder of the 1930s and in the 1940s. L.K. Sillcox, whose insistence on upgrading the Railroad Division's procedures had created the fight with Pierce Wetter, pointed out at the 1938 conference of professional divisions that 'there is an ever present threat of discontent among Divisions which feel their interests are treated as subordinate to others.'[43]. No other American engineering society had such a wide range of interests, he claimed, or depended so much on its divisions. A vice president of New York Air Brake Company and subsequently president of the Society, Sillcox had pronounced notions about technical divisions. He was particularly concerned that inefficient administration of the divisions led to mediocre papers and that commercialism, which he said could easily assume 'any proportions,' be controlled. The answer to both problems, according to Sillcox, was 'proper discrimination' in the selection of papers, but as chairman of the Committee on Professional Divisions, it proved more difficult for him to impose his rigorous standards on all the technical interest groups than it had been for him to straighten out the Railroad Division.

In any event, the technical divisions generated more papers than the Society could afford to print. In fact, in 1941 the Board of Technology recommended against more funds to stimulate papers until there was money to publish a significant number of them. It might have been imagined that the limitation of funds for publication was an easy way to regulate content, but of course things did not work so neatly. And it was not entirely clear that they should. Sol Einstein, chairman of the Machine Shop Division in 1941, argued that many valuable papers had not been published because head-

43 Minutes of the Standing Committee on Professional Divisions 5 Dec. 1938

quarters thought them too commercial. And all these interlocked problems of funding, organization, and regulation were compounded the further away from New York headquarters the technical division activities took place. R.L. Daugherty, who had been in charge of arranging technical division speakers for the March 1938 meeting of the Los Angeles section, found many of his efforts 'overturned by others who are not conversant with the local situation.'[44] The Committee on Meetings and Program in New York wanted more simultaneous sessions than the local committee thought their geographic location would support, and then organized them without the prior knowledge of the local people.

Yet even when the Society went to considerable lengths to provide support for distant groups of its members the results were not always long lasting. By 1939, for instance, the Petroleum Division was in trouble again and the Committee on Professional Divisions reported that 'many companies in the West have told their men that they should stay out of ASME and concentrate on API.'[45] And the Society's most powerful professional divisions were not exempt from these cross-currents in the technical sphere of action. Sol Einstein reported to President William A. Hanley in 1941 that, instead of rejoining ASME, many of the members who had dropped out of the Machine Shop Division during the depression had since become members of such groups as the American Society of Metals or the American Society of Tool Engineers, where they thought they received 'more professional information and education than they ever obtained from their membership in the ASME.'[46]

Einstein made two familiar but important arguments in his letter to Hanley. Whether indeed the Society needed to be reminded of it, he pointed out once again the fact that engineers tended to join the organizations they saw as most useful to them – defined by either their employer's concerns, their own interest, or a combination of the two. Just as there was a spectrum of reasons for affiliation, there was a range of organizations to join, and even in ASME's traditional areas of strength it could not automatically command the loyalty of its members. Einstein also reiterated the conventional relationship between local sections and technical divisions. Most members could only participate in the Society through local section activity, which had necessarily to cover the entire field of mechanical engineering. The divi-

44 R.L. Daugherty to the Committee on Meetings and Program, 4 Jan. 1938; Pegram Papers
45 Minutes of the Standing Committee on Professional Divisions 7 Dec. 1939
46 Council Minutes 30 Nov. 1941

sions, by contrast, had the responsibility of supplying specialized information to members, keeping them in touch with the newest developments. It was as if Einstein thought of local sections as the Society's blood supply and divisions as its nervous system.

These essential themes in the history of ASME's technical divisions, elaborated on the eve of American entry into the Second World War, were carried through to the post-war years. At the winter annual meeting in 1948, for example, in a session sponsored by the Committee on Professional Divisions and entitled 'Better Professional Divisions for a Better Society,' committee chairman W.L.H. Doyle characterized the divisions as 'the primary agencies for motivating the flow of technical information among its members.'[47] He claimed that the divisions needed more staff support and warned that, without it, specialists might transfer their allegiance to other organizations. In 1953, the Committee on Society Policy also addressed itself to the needs of divisions and their role in the organization. In a report entitled 'Developing Specialized Interests,' the committee stressed the need for 'a high standard of service' for the technical specialties and suggested an enlarged political role in the Society as a whole for leading figures in divisional work.[48]

The way professional divisions were organized into departments gave some potential for enhanced recognition of those of the Society's leaders who rose in the institution through its technical sphere of action. But not until 1966, with the creation of a new category of vice presidents, did ASME's technical side receive formal representation on Council. However, it was clear by then that the purpose of the reorganization was to give the divisions a standing in Council equal to that which the creation of regional vice presidents had given the Society's geographic representatives. Nonetheless, there were still men in the technical divisions who remained unconvinced that restructuring had provided them with adequate representation. Phillip G. Hodge, a member of the Applied Mechanics Division executive committee, argued in a 1966 letter to *Mechanical Engineering* that some elements of Council 'do not seem to be aware of the needs or even the existence of the professional divisons' and insisted they should have a still larger voice in the small circles 'where the decisions to make decisions are made.'[49] S.J. Kline, secretary of the Fluids Engineering Division, was even more blunt. In a veiled allusion to the increased attention being paid to member interests,

47 *Mechanical Engineering* (Jan. 1949) 59–62
48 Report of the Committee on Society Policy 17 June 1953
49 *Mechanical Engineering* (June 1966) 82–3

Kline emphasized ASME was 'a *technical society*,' that its primary purpose was the 'interchange of technical information.'[50] Yet as he surveyed the make-up of the new, restructured Council, he counted eleven direct votes for the regions, four indirect votes from the divisions, four technical interest votes, and one neutral vote for professional affairs. President Louis N. Rowley argued in a rejoinder that the Council should not be seen in such political terms. But anyone with Rowley's long experience in the Society knew something of its politics and realized the degree to which its social purpose and its technical purpose were competing entities as much as they were complementary systems.

Men such as Kline reached back to ASME's founding for their notion of its true *raison d'être*. Holley and Thurston had used similar language to describe the need for an institution that would make it possible for engineers to exchange information, and besides it seemed such an obvious truth. But the Society's social economy was just as much a part of the original vision – and the historical development of the professional divisions made it clear that the technical sphere of action was not characterized as much by internal logic as it might have seemed. In fact, from the beginning the divisions reflected a set of pragmatic choices and the nature of them was also evident from the outset.

50   Ibid (April 1965) 88–9

ABOVE Calvin Winsor Rice was born into a Massachusetts family which over the generations produced a number of engineers and this plaque celebrated his own long dedication to the profession.

BELOW Cornerstone-laying ceremony at the Engineering Societies Building in 1905. Frederick Winslow Taylor stands to the left of the flower girl, while behind her is Andrew Carnegie who gave the money for the building.

ABOVE The United Engineering Societies Building on West 39th Street in New York, where ASME had its headquarters from 1906 until 1960

BELOW The 'front office' in the 1920s

THE FIFTIETH ANNIVERSARY
AMERICAN SOCIETY OF
MECHANICAL ENGINEERS
ANNIVERSARY DINNER
MAYFLOWER HOTEL WASHINGTON D.C.
APRIL 8 1930.

The fiftieth anniversary medal

TOP Allegorical figures from the pageant *Control*
(left to right) Intelligence, Imagination, Mature Control, Conversion, and Finance

LEFT, TOP The St Louis section meets, with beer and hats on, in
the Ashley Street Station of the Union Electric Company.

BELOW The Society's fiftieth anniversary banquet

5

'The arts are full of reckless things'

The search for orderliness and for systematic procedures that so characterized late nineteenth-century America is usually described in political or economic terms. Historians often claim that Progressive Era politics emphasized government by city managers rather than city bosses, for instance, and that the business world also looked for a new kind of managerial expertise to take it beyond the anarchy of the robber barons. But these tendencies can also clearly be seen in engineering institutions – and they obviously derived from the same social and economic forces.

Just as in business and politics, specialized knowledge was the key to systematic technology. To create and process the particular information in which they were interested, practically all of the technical associations established in that era used the time-honored and familiar practice of learned societies – the presentation and publication of papers. It was a style that incorporated a certain concept of the diffusion of knowledge. The fruits of skilful practice, careful observation, or detailed research were shared with other experts and in that incremental fashion knowledge was advanced. That was the theory behind the technical divisions and the basis upon which they were to organize professional sessions at the Society's meetings. But in the formulation of standards, ASME found an even more potent way to translate specialized knowledge into rational conduct. Indeed, formal codes and standards of industrial practice seem the ideal expression of that drive for system that dominated American life in the years after 1880.

The extravagant pace of American technical development in the nineteenth century, like Jay Gould's fiscal sleights-of-hand, more often caused one to marvel at the results than to examine the procedures. But periodically, something happened to reveal surprising anomalies beneath the surface. Baltimore, for example, was not only among the first of the eastern seaboard cities to have a mechanics' institute, but its merchants also shrewdly backed the railroad before businessmen elsewhere realized its commercial advantages, and the city's manufacturers enjoyed a reputation for technical sophistication. Yet in 1904 its entire business district was destroyed by fire because the screw threads on its fire hydrants would not fit couplings on the hoses of fire-engines rushed from other cities and towns. The water was there in abundance, but the mechanical system failed at the most prosaic level. As James W. See had earlier pointed out to ASME members, 'the arts are full of reckless things that had better be standardized.'[1]

In some areas of technical activity, a sort of grass roots process had already brought uniformity. Burner threads in the gaslight industry, for instance, became standardized because, while makers of fixtures and makers of burners had evolved as distinct specialists, their products obviously needed to be interchangeable. It also proved natural and easy for ASME to become involved in practical standards of that sort. The process was simple. In the pipe industry, for instance, a committee composed of men familiar with the business collected information from a number of manufacturers on the dimensions of pipe flanges, and then worked out a set of standard specifications that would create uniformity of practice with the least trouble or expense to everyone. The method rested on the proposition that any standard was better than none and that widespread adoption was the key to the whole enterprise. It was also an approach suited to the penchant for specialization in American industry and to its propensity to produce a limited range of standard goods rather than, as in Europe, allowing the customer to determine such things.

Even so, the Society's members could still easily list hundreds of products that should be made to regular sizes and shapes, and they recognized the fundamental importance of that work. Oberlin Smith thought of standards as 'powerful tools for the advancement of civilization.'[2] And not only simple things needed uniform specification. William Kent had demonstrated to a roomful of mechanical engineers that each of them used a different method for testing steam-boilers. At yet another level, when Henry R. Towne tried to collect information on the procedures used in the United States to test the strength of iron and steel, he discovered there was not even a common language with which to analyze the differences. But, of still greater consequence, standards opened up important possibilities for public service, particularly in the realm of safety. Thus, there were three different kinds of pressures that pushed the Society into a vigorous program to standardize

1 James W. See 'Standards' ASME *Transactions* (1889) 573
2 Ibid 567

mechanical engineering: aesthetic notions of technological tidiness, the obvious economic benefits, and a sense of professional responsibility.

Of all the ASME standards, none more perfectly reflected those three impulses than the effort to reduce the number of steam-boiler explosions through standardized procedures for their construction, use, and inspection. Bringing order and system to the way steam-boilers were made and used was the ideal task for ASME. More of the Society's members were involved with steam-power, one way or another, than with any other single mechanical engineering subject; public safety was endangered by the frequent explosions of boilers; and the existing laws regulating them varied enormously from one jurisdiction to another. In contrast to the progress made in reducing the hazard to the travelling public of bursting steamboat boilers, no effective national methods had evolved for eliminating the threat to life and property from the large numbers of stationary boilers used to generate commercial and industrial power. Besides, business relations in the steam-boiler market-place were much disturbed by all that confusion. ASME's Boiler and Pressure Vessel Code, the first edition of which was issued in 1914, was designed to solve those problems and it soon became one of the Society's proudest accomplishments in the technical sphere of action.

The safety question has always loomed large in the history of the ASME boiler code. All the published accounts of the boiler code committee's formation and of its work have emphasized that factor. In a 1914 report, the committee identified as its primary concern 'to put an end to the killing of so many men.'[3] And the committee returned to that theme in the finished version of its first report, pointing out in the introduction that 300 to 400 people were killed each year in the United States in boiler accidents. The fact that the actual number of reported explosions in the United States had started to decrease a decade before Council appointed the committee suggests, however, that the boiler code was not simply the direct response of a public-spirited organization to a recognizable problem. Indeed, by 1914 the reported number of explosions was less than one-half of what it had been for several years at the turn of the century. The fact is there were other considerations besides safety which motivated Council in 1911 to create a 'Committee to Formulate Standard Specifications for the Construction of Steam Boilers and Other Pressure Vessels and for Care of Same in Service.'

One compelling justification for the action was that the steam-boiler industry badly need rationalization. Both producers and consumers were seriously

3 Arthur M. Greene jr *History of the ASME Boiler Code* (New York: ASME 1953) 10

troubled by a crazy quilt of regulations governing the construction and use of boilers. At the time of the committee's first report, ten states and nineteen municipalities had laws in force to regulate the construction, operation, and inspection of boilers. Not only, however, was there great variation in those rules, but also a boiler properly inspected and approved for use in one place could not usually be certified for use elsewhere. That disorder in regulatory practices created disorder in manufacturing. Producers had to maintain large inventories of boiler plate of different gauges and of all the attachments required by different rules. Furthermore, the lack of uniformity disturbed the engineering design process and inhibited the efficient development of specifications for contractual and insurance purposes. In other words, if one thinks of the steam-boiler industry as a community which included manufacturers, consumers, inspectors, and insurers, then from any point of view the situation cried out for order.

There was another element in the complex of motives that led ASME to create the boiler code committee, although its effect is more difficult to judge. Steam-power was central to the professional concerns of a large number of the Society's members and to their identification as technical specialists. They saw themselves as the authoritative body in the field and that sense of pre-eminence, enhanced by the success of ASME's standard for boiler testing, led them to a kind of technological hubris. The American Boiler Manufacturers Association had failed to adopt a set of standard specifications but as a 'commercially disinterested' learned society, ASME sought to differentiate itself from the industry's trade association. Thus, in its initial report, the committee aimed at a standard 'par excellence,' one which would do no less than create a perfect uniformity.[4]

Since the interests of so many of its members were touched by the problem, it was probably inevitable that ASME should have become involved in a boiler code. But two men, Edward Meier and John Stevens, stand out as especialy influential at the beginning. Colonel Edward D. Meier, president of the Heine Boiler Company, was an active member of the American Boiler Manufacturers Association, serving as its secretary and then as its president. In 1898, when secretary of the association, Meier had presided over a committee appointed to devise standard specifications for the construction of boilers. Trained in Germany and active in the relations of the *Verein deutscher Ingenieure* with United States engineering societies, Meier based his specifications on European experience and the best practice of American

4 Ibid

boiler makers. But even though his rules pointed more in the direction of uniform bidding and contractual arrangements than regulation, a significant number of association members saw the specifications as a threat to their own interests. Meier subsequently decided ASME might prove a better vehicle for a code the industry would find acceptable, and when he became president of the Society in 1911, he proposed it take on the task.

John A. Stevens, a consulting engineer, had been involved for many years with power generation. He had been a member of the Massachusetts Board of Boiler Rules, which in 1907 framed the first state law regulating the construction and use of steam-boilers. Called into existence by the governor after two disastrous factory boiler explosions, the Massachusetts rules incorporated a set of safety factors, construction specifications, testing methods, and procedures for certification. Moreover, the board itself had been constituted on the principle that the parties obviously interested in the problem ought to be heard. The appointees therefore consisted of representatives of boiler manufacturers, boiler users, boiler inspectors, and boiler operators.

The Massachusetts rules and the process by which they were formulated proved central to ASME's boiler code in more ways than one. In September 1911 the Council appointed Stevens chairman of its committee and also named Professor E.L. Miller of Massachusetts Institute of Technology, who had been a consultant to the Massachusetts board, as one of its members. The membership of Stevens' committee was balanced to represent interest groups, as the Massachusetts board had been, although the members were not exactly the same. Miller and R.C Carpenter, both academics, represented boiler users; William H. Boehm, a vice president of the Fidelity and Casualty Company was named to represent insurance interests; there were two representatives of boiler manufacturers, and a representative of iron and steel producers. Unlike the Massachusetts board, there was no one to represent the interests of boiler tenders, and in neither case was anyone explicitly charged to represent the public interest.

In essence, ASME's boiler code adopted the Massachusetts rules almost wholesale and then established a process for criticism and revision by the technical community. The method was not unlike that of the Society's earlier standards committees, which surveyed the field to define best practice and then suggested it as the proposed standard in a preliminary circular to interested parties. The committee's changes to the Massachusetts rules consisted mainly of reducing the allowable steam pressure in certain cases and adding design information. Otherwise, the language and organization were practically identical and in 1913 the committee sent out about 2,000 copies of its preliminary report to engineers and agencies interested in the problem. After the committee absorbed the response, it mailed out another report in 1914, which it meant to be a final draft.

A covering letter sent with this second edition reflected the committee's optimism and sense of high-minded purpose. It saw its work as an 'educational' endeavor and expected all 'well-disposed interests' to share its concern for the use of the best materials and construction methods to reduce the hazard to human life. In an effort to secure the best judgment of qualified engineers, the committee sent 2,500 copies of its report 'throughout the civilized world' and encouraged those receiving it to suggest improvements, assuring them that any 'recommendations which further protect life and property will be filed in the archives of the American Society of Mechanical Engineers against the name of anyone who suggests the same, in the spirit of 'credit where credit is due.'[5] By such an elevated and open-handed process the committee hoped for adoption of its code by the federal government and for uniform legislation across the country so 'that a boiler built for one State will be acceptable in any other State.'

As the reactions began to come in, however, the committee was shocked to discover how little its efforts were valued in some quarters. In particular, the report was opposed by those who thought it would interfere with their business interest, and they 'sowed seeds of distrust' in the minds of Council.[6] It is not difficult to identify the opposition in a general way. Some boiler makers, for instance, were against certain of the limitations on steam pressure. The 'prominent railroad officials' who protested to Council wanted a different set of inspection rules since their boilers were operated under conditions unlike those of stationary installations. There were also members of the Society who felt that as a matter of principle it should not propose legislation. But Henry Hess, a Baltimore steel manufacturer, who particularly criticized the committee for its lack of co-operation with industry and called for John Stevens' resignation, raised the important issue. What created the 'storm of protest' over the committee's report was the way it had gone about its work.

Stevens was a man of assertive personality and his deep involvement with the Massachusetts rules, which were widely recognized as the best available, had led ASME's committee to a degree of independent action that violated the essential spirit of standards formulation in America. Standards had always been the creation of the private sector, not of government, and their adoption depended upon an acceptable degree of industry participation in the development process. Thus, one of the fundamental themes in the history of

5 Ibid
6 Ibid; see also *American Machinist* (25 Feb. 1915) 345.

the boiler code was evident at the very beginning. Important interests always had to have direct and substantial representation.

Stevens' biographer suggested that the 'suspicion' the committee's work met with had partly to do with the novelty of the idea. However, Arthur Greene's history of the boiler code makes it clear that Hess's argument for throwing out the committee's work and starting the whole process anew was resolved by just the kind of industrial co-operation he claimed had been lacking. What the committee had seen as its final draft therefore became its 'preliminary report,' to be the subject of discussion at a series of meetings in September. Those hearings brought together practically every conceiveable interest group in the boiler industry. There were representatives from the Master Steam and Hot Water Fitters, the National Boiler and Radiator Manufacturers Association, the American Society for Testing Materials, the Boiler Tube Manufacturers Association, the American Society of Heating and Ventilating Engineers, the National Association of Thresher Manufacturers, the Association of American Steel Manufacturers, and the American Boiler Manufacturers Association, besides railroad interests, insurance interests, and other representatives of boiler manufacturers, users, and designers.

This remarkable array of individuals, trade associations, and technical societies was there to protect a broad set of particular interests, but its concerns were not simply selfish. It also brought detailed technical information crucial to the elaboration of a boiler code that had to be extended to cover a wider range of applications, and in greater specificity, than the Massachusetts rules envisioned. In a way that echoed the efficiency of specialization in industry, boiler tube manufacturers at the meetings worked out standard specifications for boiler tubes, while, for the first time in their history, valve manufacturers were able to agree upon uniform dimensions for safety valves. In a similar fashion, specifications for boiler materials were changed to those of the American Society for Testing Materials, which, in a piece of institutional specialization, had grown out of the campaign for standard methods of materials testing that had so interested early members of ASME. Furthermore, the close examination of various elements of boiler construction revealed other details that had yet to be resolved. But perhaps the most important result from the hearings was that they pointed out the need for additional study of particular issues, calling subcommittees into existence, and the need for a process for continuous revision.

All these criticisms were incorporated into what was called a 'Progress Report,' which the committee circulated widely and then submitted for another round of discussion at the 1914 annual meeting. An unidentified group of opponents attacked the code again at that meeting and they made a 'strong attempt' to scuttle the committee. But by that time its base of support was secure enough to resist such assaults and the stage was set for formal adoption by the Society. That required a final draft and, according to Green, the committee, together with an eighteen-man advisory group, worked thirteen hours a day, six days a week, for seven weeks to produce it.[7] Even if in a highly exaggerated form, that effort illustrated another significant element of the boiler code's history. The work took considerable time and volunteers to the committee had to have the solid backing of their employers. Or, to put it another way, large firms with a direct interest in the boiler code could most easily afford to have their engineers on the committee.

The role of David S. Jacobus in the early history of the boiler code illustrates what a senior engineer of a large and influential company could accomplish. Jacobus left active teaching at Stevens Institute in 1906 to join Babcock and Wilcox Company in the capacity of 'Advisory Engineer.' Both of the firm's principal officers were among the founding members of ASME and George Babcock served as president in 1886–87. By the time Jacobus joined the company, he already had a reputation as one of the country's leading boiler experts and he was also distinguished within the Society both for his extensive participation in professional sessions and for the generous amount of time he gave to committee work. These contributions were rewarded by a vice presidency in 1903–05 and election to the presidency in 1916.

An important part of Jacobus's new job was to keep abreast of political developments that touched the company's interests. Thus, in 1909, when the Massachusetts legislature held a hearing on the proposals of its Board of Boiler Rules, Jacobus went to Boston. The members of both the legislative committee and of the board expected his presence to mean opposition, but in fact Jacobus spoke in favor of the rules and this support from one of the industry's leading firms apparently impressed the legislative committee. According to those who knew him Jacobus 'knew how to deal with people,' and Arthur M. Greene claimed it was his timely intervention at the 1914 spring meeting that persuaded Henry Hess to drop his attack.[8] It was also Jacobus who came up with the idea of public meetings to insure industry participation in the preparation of the code. He was appointed, as a representative of boiler manufacturing interests, to the advisory committee that helped draft the final version, and in 1915 to the committee itself. From 1918 to 1936 Jacobus was one of the committee's vice chairmen and when

7 Greene *History* 14
8 David S. Jacobus, Biographical Files; ASME Archives, New York

Fred Low died in 1936, he became chairman, which he remained until 1941. For over three decades, then, Jacobus maintained the Babcock and Wilcox 'seat' on the committee, a term its members commonly used to express the fact that major firms in the industry were always represented.

It has long been a truism that industrial support was necessary for the work of the technical divisions. In the boiler code it proved absolutely crucial. Over the years, the main committee and all its subcommittees absorbed an enormous amount of labor. And, besides the considerable time spent at meetings, men who have been active in the committee always recall hours of homework. But, in addition to the aid of highly trained and expensive people, industry has also contributed research and testing activities of substantial value. Jacobus opposed the idea of appealing to business for direct cash support for the committee's work, but several firms such as Babcock and Wilcox, and Combustion Engineering, indirectly provided incalculable amounts of financial assistance.

To the leading figures in the boiler code, there was no contradiction between their claim that it served the public interest and the fact that it was framed and financed by the boiler industry. Herbert Spencer's argument that self-interest was the foundation for scientifically correct conduct continued to provide the men of the boiler code committee with a rationale for their work long after Robert Thurston had applied that philosophy to ASME's purposes. Indeed, their own experiences encompassed more than one case in which Babcock and Wilcox, or Taylor Forge, or some other firm had given the committee useful data from research undertaken because of a commercial interest in the problem. They also felt a sense of professional responsibility that encouraged a broader concept of interest. As Arthur Greene said at the testimonial dinner for Jacobus on his retirement in 1941 as chairman of the committee, 'politicians don't understand that a man can rise above his selfish interest.'[9]

For all their concern to portray the code as a public-spirited enterprise motivated only by considerations of safety, these men certainly knew about the economic power of specialized knowledge and its effectiveness as a weapon in the market-place. They also knew that some of ASME's members often used or tried to use its programs and activities to further the commercial concerns of their employers. But there were several factors that helped to prevent the debasement of the code. In the first place, committee members were sensitive to the fact that industry had a direct interest in their work,

9 Ibid

and, as one long-time member put it, they watched each other 'pretty closely.' Taken by itself, that sort of internal policing might not have proved very effective, although men who worked on the code for any length of time tended to identify themselves with it, especially as it became known as one of the country's outstanding examples of a safety standard.

Besides these personal and subjective factors, however, there were some practical reasons prompting the boiler code committee to strive for a reputation of great objectivity. For one thing, the code helped settle the domestic market-place of the boiler industry – and probably to the advantage of larger firms – just as the conservation movement's insistence upon technical expertise tended to favor bigger companies in the forestry industry. Thus major boiler firms had an important stake in a widely adopted code. Not only did it serve to rationalize the design process and contractual relations between producer and purchaser, but it also brought system and order to boiler inspection, certification, and the insurance business. Manufacturers, for instance, had previously to produce different boilers for different jurisdictions, and at the opposite end of the system inspectors had had to be licensed in different jurisdictions, all with their own requirements.

Although no one ever talked much about them, there were some other advantages the boiler code may have given the industry. For instance, it perhaps helped with labor problems. A manpower shortage in a particular craft could sometimes be solved without recourse to higher wages by amending the code. That strategy could also have lent itself to the avoidance of troublesome jurisdictional lines between crafts. And just as the code helped order the domestic market-place, it tended to exclude foreign products, although the Society denied that charge when it was made.

The obvious wisdom of a widely used standard set of specifications was not lost on the industry and, simultaneously with the formulation of the first boiler code, the American Uniform Boiler Law Society was created. Thomas Durham of the American Boiler Manufacturers Association led the movement and in 1915 he became the chairman of the new society's administrative council. Since the boiler law society sprang from the industry, its administrative council consisted of representatives from trade associations and the manufacturers of various types of boilers, from the insurance industry, from the producers of boiler-making materials, and from the National Electric Light Association, which spoke for the electric utility industry. In other words, the same important interests that staffed boiler code committees were those that pushed for the code's widespread adoption.

The final step in this co-ordinated effort to organize a sometimes dangerous and certainly disorderly industry was to create a standardized inspection system. ASME's boiler code committee had recommended a set of rules for the qualification of inspectors in 1918 and the following year C.O. Meyers, in a paper before the American Boiler Manufacturers Association, also called for uniform procedures in the licensing of boiler inspectors. His notion was that both organizations should work together through a joint committee to frame standard rules, and at a meeting of that conference committee, in September 1919, the basic outlines of a National Board of Boiler and Pressure Vessel Inspectors were drafted. With funding from the American Boiler Manufacturers Association and the insurance industry, and temporary housing from the American Uniform Boiler Laws Society, the board began to administer uniform standards for boiler inspectors working within jurisdictions that had adopted the ASME boiler code.

There seems, then, little doubt that out of a pragmatic combination of idealism and self-interest, much in the way Thurston imagined, most of the members of the boiler code committee, and of its subcommittees as the scope of the enterprise expanded, worked hard to produce a technologically rational document that aimed to harmonize the financial ambitions of industry with public concerns for safety. The heart of the matter was a set of procedures designed to achieve that practical equilibrium. The procedures were based, in the first instance, on the already well established practice that those in a given industry who best knew its technological and economic requirements should play the major role in defining its standards. That connection was essential in a country where government did practically nothing toward the creation of standards, and where their adoption depended on acceptance by the private sector. Furthermore, the idea that the representative interests included consumers as well as producers carried with it a notion of checks and balances, which was enhanced by the subsequent addition to code committees of supposedly neutral engineers to represent the public. Still another important procedure that aimed to insure open access to the formulation process was the publicizing of the committee's work and meetings. While they did not call it that, what the committee had in mind was a concept similar to due process, which to Morris L. Cooke stood out in sharp contrast to the secretive ways of Council. Why shouldn't the Society have a code of ethics, he asked Calvin Rice, as 'rigorous' as the boiler code?[10]

A set of generally effective procedures, a sense of professional purpose, and the active interest of the industry's major firms all combined to give ASME's boiler code the kind of reputation its creators hoped for, and it was eventually adopted as law in most of the states of the union, and sections of it were also

10 Morris L. Cooke to Calvin W. Rice, 8 Sept. 1916; Cooke Papers

adopted in the Canadian provinces, Australia, and Great Britain. It proved a model voluntary standard and was often used as an example of the ability of the private sector to generate standards that served the public interest.

Yet, historically, the boiler code is also a model of the constraints typical in the Society's technical sphere of action. The committee's dependence on trade associations for advice, for instance, yielded both useful information and industry acceptance, but it also exposed the code to the dangers of sub-version. An example of just that sort of possibility occurred in 1937. P.C. McAbee, owner of Carbonic Gas Equipment Company, of New York, claimed that a 1936 addendum to the unfired pressure vessel code was the fruit of a conspiracy intended to drive him out of business. McAbee's firm manufactured dry ice converters that he delivered empty to his customers and he had on hand a large number of orders for those containers, which he could not produce by the modified rules without substantial loss. McAbee also charged that his competitors, the principal producers of liquid carbon dioxide, 'virtually controlled the carbonic gas industry' and also dominated the Compressed Gas Manufacturers' Association.[11] The new rules had been formulated at the request of that association and its members made up a majority of the 'special' committee which drafted them.

The new rules required containers of the kind McAbee's firm produced to sustain an impact test, and this, according to his attorney, was only possible through the use of a steel 'not obtainable in the American market as a stock item.' But apart from the alleged unreasonableness of the technical difficul-ties posed by the revised specifications, they were not applied logically or evenly. Liquid gas producers, whose interstate shipments were covered by regulations of the Interstate Commerce Commission instead of the ASME code, regularly delivered gas at higher pressures than that in the stationary converters of McAbee's customers. Furthermore, his attorney claimed, the liquid gas producers changed their argument after they lost a legal action McAbee had brought against them under the Sherman anti-trust laws; prior to that, the Compressed Gas Association had not argued for impact tests of solid ice converters in its brief to the New York City Fire Department or to the Interstate Commerce Commission. In other words, the situation looked completely contrived to McAbee, and his lawyer advised ASME Secretary Clarence Davies that his client saw little alternative than another legal suit.

In his own memorandum of the events surrounding the case, Davies got right to the heart of things. The central question was whether 'the Boiler

11 Minutes of the Executive Committee of Council 3 Oct. 1937

Code Committee was not being made a tool of a group of manufacturers of liquid carbon dioxide in their competition with distributors of solid carbon dioxide.[12] After considerable discussion with the Society's legal counsel and with the principal actors in the drama – C.W. Obert, long-time secretary of the boiler code committee and a member of the special committee; its chairman, H. LeRoy Whitney of the M.W. Kellog Company; representatives of the Compressed Gas Association; and David Jacobus and H.E. Aldrich, chairman and vice chairman of the boiler code committee – the Society's leaders decided that a 'revision procedure' was in order, and that anyone interested in the issue be invited to participate. In that relatively quiet fashion, the boiler code committee interpreted the new rules to exclude McAbee's converters from impact testing and avoided what might have been an embarrassing and costly affair.

But here lay the dilemma: industry participation was essential to the code, and that was the way most members of the committee thought things ought to be done; yet it automatically created the suspicion, if not the reality, of a small group of men whose primary concern was the interest of the industries they represented. And besides the actual fact of representation, the circumstances of the committee's operation gave further impetus to the suggestion that it was beyond the usual restrictions. For instance, to avoid the normal requirement that the personnel of standing committees be rotated, Council made the boiler code committee a special committee so that its members could be reappointed each year. Thus one of its outstanding features became the long tenure of its members, many of whom, like Jacobus, served over three decades.

The committee also gave the appearance of a small tightly connected group because the same men were involved in so many related activities. For example, members of the committee played important roles in the American Boiler Manufacturers Association, the Uniform Boiler Law Society, and the National Board of Boiler and Pressure Vessel Inspectors. Furthermore, at a personal level the committee often acted the way clubs do. Its members usually occupied senior positions in industry, they met frequently, formed long-lasting friendships, and generally shared the same social, economic, and political viewpoints. And in the way such groups work, they occasionally took care of colleagues in times of difficulty. Its older members remember more than one case in which that happened. For instance, when the depression of the 1930s forced a cut-back in ASME staff, David Jacobus found C.W. Obert a job with Union Carbide. There were other ways in which the boiler

12 Ibid

code committee was like a club. For years it was almost entirely concerned with steam-power. Young engineers coming into code work for the first time in the late 1920s, and more interested in unfired pressure vessels, felt as if they were entering an exclusive and traditional preserve, while members from the chemical and petroleum industries found the position of chairman all but unobtainable.

This close co-operation, with such personal, professional, and business implications to it, created in ASME's boiler code committee an institutional device of great power. Although basically conservative in its tendencies, in time the committee's work encompassed new technologies for the fabrication of fired and unfired pressure vessels, new ranges of temperatures and pressures, and the new materials they required. Specifications for high-pressure gas and oil pipelines also became part of the code, and after the Second World War the committee took on the task of formulating standards for nuclear power generation – a job with even greater significance than anything that had gone before.

The essential nature of the boiler code committee remained unchanged, however, as did the context in which it operated. Over the years, many of its long-term members gave a great deal of time and energy to code work, out of a genuine sense of professional obligation. But the fact that in America a direct connection to industry continued to be an active element in the formulation of standards meant not only that the code was vulnerable to attack on the grounds of self-interest, but also that it was sometimes guilty of the charge.

6

Technological Confidence during the Great Depression

To the men of the American Society of Mechanical Engineers, proof of the value of the organization lay in America's remarkable technological triumphs. They could see the connection in several different ways. Most obviously, there were members famous for their accomplishments. George Westinghouse, for instance, ASME's twenty-ninth president, reached the status of folk hero for his inventiveness and the giant industrial firm he created. There were others like him, such as George Babcock, Ambrose Swasey, and Charles Schwab, who, although not as well known, were held in great esteem in the mechanical community for their technical achievements and business enterprise.

The Society's meetings and *Transactions* suggested another way in which it directly aided the advance of the mechanic arts in America. Outstanding papers dramatized the importance of professional publications as a means of adding to knowledge. Henry R. Towne's suggestive sketch, 'The Engineer as Economist,' for instance, and Frederick Winslow Taylor's classic exposition of a system for piece-rate wage payments established in the new field of scientific management a mechanical engineering specialty of enormous future importance. In a less spectacular but equally long-lasting fashion, Robert H. Thurston's 1893 paper on engineering education formulated the philosophy of American technical education still dominant today.

ASME's research projects pointed out yet another way in which it served the cause of America's technical progress. There had always been hopes that the Society would be able to encourage original investigations into fundamental engineering problems, but until the creation of the Engineering Foundation in 1915, through the benefactions of Ambrose Swasey, there was no endowment for such purposes. Swasey's hopes to stimulate other and larger contributions, like Herbert Hoover's ambitions in the 1920s for a great industry-financed campaign of applied research, were never realized, although, through a 'good natured apportionment' of the interest on the Engineering Foundation's capital, the Society financed a few research projects. Most of the more than thirty co-operative investigations under way at the beginning of the 1930s, however, depended on industrial contributions and the research committee was accordingly 'made up largely of men sympathetic to research but of an executive type,' who were also chosen to represent major branches of mechanical industry and the principal sections of the country.[1] By such strategies and by identifying specific, practical problems, ASME's Committee on Research was able to organize a number of useful investigations. A project at Massachusetts Institute of Technology to study the properties of high pressure steam, for example, was carried on with funds from the National Bureau of Standards and money George A. Orrok, a New York Edison Company engineer, so successfully raised from the utility industry.

In a more prosaic but equally apparent fashion, particularly to men who recognized that engineering advance was a slow and incremental process, the Society's leaders could measure its contributions to the country's technical life through the increasing number of local sections, the creation of new professional divisions and student chapters, and a membership growth through the 1920s that accelerated at a rising rate every year. Much of that sense of accomplishment, the conviction that engineering and American civilization were inseparably linked and that mechanical technology was at the heart of the nation's industrial pre-eminence, was reflected in the theme chosen for ASME's fiftieth anniversary. Theirs was the 'Society of the Industries,' mechanical engineers claimed, and they proudly associated themselves with the astonishing bounty that poured out of America's highly mechanized, 'scientifically' managed factories.

The Great Depression of the 1930s cast all that self-confident optimism into doubt, and not simply because the economic calamity indicated faults in the free enterprise system – the depression was also a technological crisis of the first order. Never before had Americans felt such uncertainty about their traditional mechanical ingenuity and its supposedly beneficial effects. Indeed, there seemed to be plenty of evidence that problems which appeared economic were actually technological. Machines that saved labor created unemployment; mass production meant over-production. The Machine Civilization debate of the 1920s had already raised troubling questions about the social effects of mechanization, but after the stock market crash of

1   C.B. LePage to Ralph E. Flanders, 13 Nov. 1935; Flanders Papers, Syracruse University

October 1929, the very qualities of the modern age that mechanical engineers celebrated as their particular contribution took on the character of aberrations which exaggerated the swing and severity of normal business cycles and turned them into catastrophe.

Sociologist Clarence M. Case echoed the attitude of many depression era critics when he claimed that what the country needed was 'social engineering not mechanical engineering.'[2] His remark only pointed up what the Society's prominent men already felt; more than for any other engineering institution, the depression was a special problem to ASME. Because it claimed industrial production as its particular area of expertise, the Society risked having to shoulder the blame for over-production. Yet, because of its professional qualifications in industrial management, the opportunity lay open for it to play an especially prominent role in the recovery process. And not only did ASME's leaders have to defend their profession against a rising tide of opposition to technology. The Society itself was threatened by serious loss of membership – unemployed engineers found themselves unable to pay their dues, while hard times led still others to wonder about headquarters mismanagement and the dominant influence of New Yorkers in the organization's affairs.

There was practically no one within ASME who did not have some conviction about the causes of the depression. George Stetson, editor of *Mechanical Engineering*, recalled 'being swamped with manuscripts and letters' on the subject.[3] But chief among those seeking to defend mechanization by an economic analysis of the calamity was Ralph Flanders, who had so forcefully argued the ethical case for engineering in Charles A. Beard's *Toward Civilization*. A machine tool manufacturer and subsequently United States senator from Vermont, Flanders was convinced that the crisis resulted from problems of distribution rather than of production, and from excessive conservatism in financial circles. The issue was not scarcity; 'there can be plenty for all in a mechanized society,' he argued, if the economic system could only learn how 'to direct and distribute the flood of goods that engineering skill can make available.' But instead of the straightforward and efficient organization that characterized productive units, distribution facilities were clogged with 'lawyers, the advertising men, the brokers, the vast sales organization, the holding companies, and so on without end.'[4] Flanders claimed that in a

2 As quoted in William E. Akin *Technocracy and the American Dream: The Technocrat Movement, 1900–1941* (Berkeley: University of California Press 1977) 165
3 George A. Stetson to Ralph E. Flanders, 2 July 1935; Flanders Papers
4 Charles A. Beard ed *Toward Civilization* (New York: Longmans, Green 1930) 26–7

similar way investment capital was also ineffectively employed. Calling up a wartime image, he spoke of 'slacker millions which are towering in the financial institutions of New York and other great cities,' and which, for want of confidence, remained unavailable to help solve the unemployment problem.[5]

The irritation Flanders expressed at supernumeraries in the distribution system and capitalist cowards who hid from battle reflected some of the dilemma felt by engineers in general. Not only were they held responsible for the millions of unemployed – as if they had all been displaced by machinery; but there were many critics who also argued for a halt in engineering applications, on the theory that *social* invention needed to catch up. From one quarter, Walter Lippmann assailed an engineering concept of government which in its extreme forms constituted 'a monstrous blasphemy against life itself.'[6] From the other side, secretary of agriculture Henry A. Wallace charged engineers with a lack of regard for the social implications of their work, for economic conservatism, and for a ruthlessly competitive approach to human relations. 'No great harm would be done,' he said, 'if a certain amount of technical efficiency in engineering were traded for a somewhat broader base in general culture.'[7]

These assaults proved particularly galling since, among professional groups, engineers were hit hardest by unemployment because of retrenchment in manufacturing and the capital goods industries. The depression also served to remind them of the anomalous nature of their calling. America's industrial capability depended on their specialized knowledge, yet they tended to occupy subordinate positions in the industrial hierarchy. Furthermore, while it seemed evident that the same combination of technical and organizational skills that had wrought such miracles of production were precisely those needed to solve the depression's problems, their lack of real power provided another reminder of an often marginal status. As Arthur V. Sheridan put it in a letter to Ralph Flanders, 'The engineer has created modern society but has not been permitted to administer it.'[8]

The engineers whose active commitment of time and energy to ASME elevated them to high offices were ambitious men, both for themselves and for their profession. It was obviously in Ralph Flanders' business interest to argue against popular attitudes or government policies that might inhibit

5 Ralph E. Flanders to Clarence E. Davies, 23 Sept. 1931; Flanders Papers
6 Lippmann's remarks appeared in the *New York Herald-Tribune*, 20 Dec. 1935.
7 Henry A. Wallace 'The Social Advantages and Disadvantages of the Engineering-Scientific Approach to Civilization' *Science* (5 Jan. 1934) 3
8 Arthur V. Sheridan to Ralph E. Flanders, 22 March 1932; Flanders Papers

sales of the machine tools his firm produced, and he used his position in the Society, just as he did in the National Machine Tool Builders Association and in his work for the NRA, to argue that the best solution to the depression was more mechanization, not less. But, along with so many other engineers in the 1930s, Flanders was also deeply convinced that mechanical advancement and industrial productivity were crucial to the country's revival. Such men were equally certain that their training, methods, and experience made them the most capable of all groups to deal with the crisis. Thus, out of a mixture of self-interest and professional conviction – the combination that so often inspired the Society's actions, ASME's leaders imagined themselves especially responsible for solving the depression's problems.

The ideas of planning and the rational employment of resources were congenial to engineers, many of whom could remember industrial mobilization and the nationalization of railroads during the world war. The depression seemed the same kind of crisis, but the first order of business was to determine causes. That in itself promised to be a complex undertaking. Furthermore, the enterprise called for a broad political foundation if its findings were to be persuasive. The American Engineering Council, which in its early years had served as an instrument of progressive reform, seemed just the vehicle for an investigation into the real causes of the depression. The topic was reminiscent of its original willingness to tackle important public issues – the elimination of waste in industry, for example – and it claimed to speak broadly for engineers throughout America. In fact, the AEC had proven little more successful in pulling together the country's engineers than any other 'unity' organization, and it had long since lost much of its crusading zeal. But, for Ralph Flanders and the activists in ASME's Management Division who wanted to conduct an investigation into the depression's causes and who looked to private philanthropy for its funding, the AEC provided just the kind of umbrella they needed.

At the Council's meeting in January 1931, Flanders presented a report that he, Lawrence W. Wallace, and Leon P. Alford had drafted. All three were interested in scientific management. Wallace, long-time secretary of the AEC, had edited its famous study of industrial waste, while Alford wrote the life of H.L. Gantt, Taylor's most prominent disciple, in ASME's biography series. Their mutual concern for efficiency, whether in industry or society, thus led them to collaborate in a management-style analysis of the depression. Entitled 'Balancing the Forces of Consumption, Production and Distribution,' their report aimed to identify the essential questions needing further study and the appropriate method for the investigation. There was nothing subtle in it. Engineers did not usually think about restricting output to solve the problem of over-production. They wondered rather how to increase consumption, how to obtain the efficiency of sophisticated technology without an excessive investment in capital equipment or further unemployment, and whether more technical research and development would not be a better long-term solution than public works expenditures – although in the latter they also saw the need for their organizational skills. And the method that Flanders and his colleagues advocated for the investigation was equally a reflection of engineering experience. Much in the way the Society's committees had generated standards for industrial practice, Flanders' group proposed integrating the existing studies being done by government, business, trade associations, technical societies, and other private research organizations. Just as that kind of systematic processing of information had proved so valuable in promoting the prosperity of the '20s, so it would provide a way out of the depression. At a time when everyone had a solution to that problem, it remained only to find a benefactor willing to give engineers a chance where economists and politicians had failed.

For a brief moment in 1931, it looked as if that patron had been found from within ASME's own ranks. Roy V. Wright, its president that year and a man deeply concerned with the question of social responsibility, attended the Southwest Power Conference in Kansas City, representing the Society. As Wright recounted the story afterwards, a reporter approached him for a copy of his speech, prior to its delivery, in order to make a deadline. However, since Wright had planned to speak from rough notes, he could give only a general idea of the points he meant to make. Among other issues, he planned to address the question of technological unemployment and to acknowledge the engineers' share of responsibility for it, as a way of arguing for the important part they might play in resolving the problem if funding could be found for the AEC study Flanders' group had recommended. The newspaper, however, carried a banner headline the next day, 'Engineer Places Slump Blame on Own Profession,' making it appear that a nationally prominent engineering spokesman accepted the proposition that the calamity had been caused by technologically created over-production.

The newspaper account of Wright's speech raised a storm among engineers, who were already sensitive to criticism for their part in bringing about the depression. But the most surprising result was a long telegram to Wright sent the next day from the Westchester Country Club in New York by Henry L. Doherty, an ASME member, one of the organizers of the Power Division, and a man made wealthy from his interests in the private utility industry. Doherty offered to provide $500,000 for the investigation Wright had

described, stipulating only that it be co-ordinated with the Kansas City
Chamber of Commerce, and he promised to raise an additional five million
dollars 'if you really have any sound ideas.'[9] Doherty's proposal excited a
great deal of attention and hundreds of people wrote to him with remedies of
their own. But in practice, his conditions proved more difficult to meet than
they had first seemed. In a subsequent letter to Calvin Rice, Doherty said 'I
cannot believe that Dr. Wright has any plan whatever,' and he hinted darkly
at a Kansas City newspaper war in which he was involved that suggested
other complications.[10] But the greatest impediment of all, as the affair
unfolded, was his insistence that any plan must be immediately effective. No
one could have supplied that.

Eccentric and short-lived as it was, the Doherty episode illuminates the
difficulties ASME's leaders had with the charge that engineers were responsible for technological unemployment. Already sensitive to public criticism
on that subject, some of the Society's prominent members opposed any discussion of the issue, or at least any that might sound like an admission of
guilt, and cited the press's treatment of Wright's speech as an example of the
hazards to even a sincere effort to address the question. For instance, Ely C.
Hutchinson, editor of *Power*, claimed he was 'out of sympathy with any
willingness on the part of engineers to accept any blame of the personal and
sociological nature, which some people are endeavoring to place upon the
engineering profession for having advanced technology.'[11] Although he
would later change his mind about the relation between mechanization and
unemployment, Hutchinson exemplified the defensive reaction of many
engineers to that issue. Yet ASME's spokesmen could no more avoid the
subject of technological unemployment than they could the defense of their
profession. The newspapers were obsessed with the subject and by 1931
there were public hearings, at practically every level of government in the
United States, on the problem of unemployment caused by 'technological
advance.' The Society felt compelled to respond. And once it was clear that no
understanding with Doherty was possible, it became important to announce
the fact in a way that did not imply ASME had failed because engineers truly
were responsible for the depression's jobless hordes. Finally, at a more
personal level, the episode was unsettling because it encouraged so many
unemployed engineers to hope they might find work in the investigation.

9 Henry L. Doherty to Roy V. Wright, 9 Sept. 1931; Flanders Papers
10 Henry L. Doherty to Calvin W. Rice, 24 Sept. 1931; Flanders Papers
11 Ely C. Hutchinson to Ralph E. Flanders, 12 Oct. 1931; Flanders Papers

For Ralph Flanders, who had committed himself to a public role in the analysis of the depression, the next step was to seek alternative sources of funding. He already had cause to believe that the Falk Foundation, a Pittsburgh philanthropy, would entertain a proposal along the lines of the one he, Alford, and Wallace had put together for Doherty, so he submitted it in a revised form. The essence of the plan was the synthesis of existing studies to identify sub-sets of the general problem amenable to solution through the application of engineering experience combined with specialized knowledge. The directness of the idea gave it an attractive quality at a time when there were any number of studies and proposals already under way. The problem was not a lack of research – as the Society's brief put it, 'the mass of research done and doing is appalling.' What Flanders proposed was a Committee on Economic and Social Progress, with its own paid secretariat, that would survey and integrate existing data, hear expert testimony, and frame practical recommendations. The group might take on, for example, the question of public works, still an embryonic concept in 1931. It might also look at the organization of manpower for productive purposes – a problem that management-oriented engineers felt especially competent to address. After that, Flanders suggested to the Falk Foundation's director, the committee might 'be entrusted with other elements of the economic problem in their order – such as technological unemployment, balance between demand and capital expansion, etc.'[12]

Throughout the summer and fall of 1931 Flanders, Wallace, and Alford polished their proposal to the Falk Foundation and carried on an active correspondence with its officials in anticipation of a favorable response. After all, the foundation had declared its interest in economic matters and, because of various magazine articles he had written on the depression, had approached Flanders for his advice on the ways a philanthropic organization might use its funds to help solve the nation's economic problems. In November, however, the ASME group got the bad news that the foundation had decided not to support the proposal. There is nothing in Flanders' correspondence to suggest the foundation's reasons for its decision, but Wallace thought the AEC project was turned down because a similar proposal, advanced by Alfred D. Flinn, executive secretary of the Engineering Foundation, had confused the situation.

Under the best of circumstances, engineers had a hard time convincing fellow Americans of their importance as a profession. The Engineering

12   Ralph E. Flanders to J. Steele Gow, 2 May 1931; Flanders Papers

Foundation's competing proposal to carry out an investigation into the
depression's causes only made that difficult task of public relations all the
more troublesome, just when the AEC was itself looking for a new prominence and effectiveness as the agency that spoke for all of America's engineers in this time of crisis.

Technocracy – a depression-born movement to reorganize the country's production and distribution systems – posed another threat of the same kind,
but with even more distressing implications, especially for ASME. In the first
place, and most damaging of all, Technocracy's spokesmen, Howard Scott
and Walter Rautenstrauch, claimed to have scientific evidence that the
depression had been brought about by technological unemployment. An
elaborate 'Energy Survey of North America' purported to document the
extent to which machines had displaced labor, and it immediately attracted
nation-wide attention. But it was not simply that the Society was troubled
because Technocracy further complicated engineering's public image, or that
the new movement undermined Ralph Flanders' campaign to identify the
economic causes of the depression, although both were central concerns.
Technocracy also had elements of internal subversion to it and ASME's
leaders felt betrayed.

    Howard Scott, later so personally identified with the movement, was rela-
tively easy to dismiss. His background and credentials were uncertain and his
manner idiosyncratic. But Walter Rautenstrauch, who had been a member of
ASME's Council, was another matter. In 1932, when Technocracy first burst
on the scene, Rautenstrauch's status as head of Columbia University's department of industrial engineering gave the movement its intellectual authority.
His standing in the university persuaded its president, Nicholas Murray
Butler, to endorse the movement, which also attracted other prominent
engineers. But Calvin Rice, who had spent his life promoting the advance-
ment of the engineering profession, was scandalized by the emergence of
Scott's Committee on Technocracy research group under the auspices of
Columbia University. He thought the whole thing 'disgraceful' and told
Rautenstrauch so. Even more distressing, he wrote to Ralph Flanders,

Ely Hutchinson and Mr. Carmody, Editor of Factory and Industrial Management
went up to Columbia last Thursday and instead of protesting they came away much
impressed with the facts that have been collected. Mr. Carmody says it is up to the
engineers to face the facts such as these. First, we as engineers have been making no
systematic provision for those put out of employment by our technological development. Second, we have no constructive program to put to work the 11,000,000

unemployed and, third, have no plan even ultimately of putting them all to work. We are lying down and leaving it to the social workers. He asks, 'What are the *engineers* going to do about it?'[13]

Rice was particularly aroused by the statement on technological unemployment that Scott issued in the form of a press release and which received wide coverage in the New York newspapers. L.W. Wallace was angry because he had been spending so much of his own time and energy over the past three years to come up with engineering solutions to the unemployment problem. And what made Scott's charges especially offensive was that Rautenstrauch, Hutchinson, and Carmody gave them credence. He was amazed to learn, Wallace wrote Rice, that 'Dr. Rautenstrauch was a party to what appears to be a rather superficial Bolshevic undertaking,' and that he had not instead first worked through ASME to discover the activities engineering agencies had under way to deal with these issues.[14] He himself had spent four months as a member of a Senate committee studying the question of technological unemployment and knew that Scott's claims were factually incorrect, but neither Rautenstrauch nor Carmody, he argued, appeared aware of the report of that committee, or of the report of Flanders' committee, or, in fact, of the work of any of fifty other committees in the engineering profession dealing with the unemployment problem.

The Technocracy movement highlighted the difficult situation in which ASME found itself. More than any other group within the profession, mechanical engineers felt the sting of criticism for technological unemployment and imagined themselves most competent to deal with the industrial system's imbalances. Yet without the ability to focus public attention or to command political power, there was little the Society could do either to defend mechanical engineering or to advance its own remedies for the crisis. It proved relatively easy to persuade Rautenstrauch of the error of supporting Scott, whose own platform excesses might well have done that anyhow, and Rice reported to Flanders, in December 1932, that Rautenstrauch had become 'disillusioned.' But what the Society could do by way of a positive response to Technocracy's claim that capitalist mechanization had created so much, and seemingly permanent, unemployment, was less simple to decide. Rice proposed 'a quiet, dignified, article telling the wonderful contributions to civilization of science, engineering and invention.'[15] But Ralph Flanders

13   Calvin W. Rice to Ralph E. Flanders, 6 Sept. 1932; Flanders Papers
14   L.W. Wallace to Calvin W. Rice, 8 Sept. 1932; Flanders Papers
15   Calvin W. Rice to Ralph E. Flanders, 17 Dec. 1932; Flanders Papers

was less optimistic about the value of yet another *Mechanical Engineering* piece in that vein. Young engineers were more concerned with present problems than past accomplishments, he replied, arguing also that the Society had to think that way, too.

Both Flanders and Wallace still hoped to find some sort of funding for their proposal, but they also looked for other ways to make the AEC an effective mechanism for engineers to use in addressing the technological issues the depression had raised and to give technical men an opportunity to use their skills in solving its problems. For example, through Wallace and Flanders, the Council worked closely, but behind the scenes, with Senator Robert LaFollette jr on drafting a bill for the administration of public works. And President Hoover, one of the principal moving spirits behind the creation of the Council in the first place, used Wallace as a sounding board for his ideas on a permanent structure for public works. But Franklin D. Roosevelt proved much less interested in the advice of engineers, and, in any event, the AEC tended to be of somewhat limited utility since it was so difficult for the profession to agree upon its collective role in the depression and its position on the political solutions advanced to deal with it. The historical combination that had earlier generated such concern for social responsibility – Progressive Era notions of reform through expertise and Scientific Management's concept of an objective engineering professionalism in the interest of optimal productivity – lacked the dramatic impact it once had. Indeed, the utopian vision of social justice through abundance just seemed ironic in the 1930s.

Yet it was hard for ASME engineers to give up that central ideological element. They opposed such anti-technological remedies for the depression as a moratorium on invention, restrictions on the introduction of new machines, or cut-backs in productivity, not only because of the economic implications for engineers, but also because they were so fundamentally committed to the view that material plenty could solve social problems. Instead of a world in which one class benefitted only at the expense of another, mechanical engineers imagined unlimited potential wealth and thought that by eliminating the historical conditions of human servitude they might resolve class conflict without revolutionary violence. And that spectre was not so remote. Those in the Society most actively concerned to connect it to the current crisis – Flanders, Wallace, George Stetson, and Clarence E. Davies, for example – noticed the appeal Norman Thomas's socialistic ideas had for young engineers, and recognized the need to improve their own understanding of Marxism as well as the importance of economic study to engineers generally. As most of his colleagues did, Ralph Flanders approached the study of economics looking for some means of regulating the balance of production, distribution, and consumption. The metaphor that most often came to mind was that of a steam-engine governor which prevented wild and uncontrolled behavior; but apart from pointing out the need for social engineering – a phrase meant to suggest the conjunction of social problems and engineering skill – few of ASME's leaders advocated any radical alteration in the status quo.

That sense of order also determined the Society's response to Technocracy. In a confidential memorandum he circulated among a dozen of ASME's inner circle, following a meeting several of them had with Scott, Flanders opposed a direct assault on the movement. Instead, in the interest of economic recovery and 'the prestige of our profession,' he urged 'attacking the economic problem itself, directly, immediately and publicly.'[16] Thus, Flanders came back to the idea of something like the proposed AEC study, an investigation that would employ the combined intellectual resources of the mechanical community. That notion of using engineering societies and trade associations to analyse imbalances in the productive system shared some features of a plan advanced by Gerard Swope, president of General Electric, and of the NRA when it was launched. Like those approaches, the Society's response to the economic issues raised by the depression was essentially conservative, and left to businessmen the responsibility for finding an equilibrium between production and consumption.

Still, Scott's charge that an uncontrolled technology had created the depression's massive unemployment, and the widespread attention his claim attracted, called for an explicit response to that particular issue. What seemed most appropriate to George Stetson, editor of *Mechanical Engineering*, was 'a series of articles which would develop a backfire of constructive comment on the very problems that technocracy has brought up.' Furthermore, Stetson thought Dexter Kimball's forthcoming address at the annual meeting of the American Association for the Advancement of Science would demonstrate, 'without attempting to ignore or minimize the problems raised by technological progress,' just the sort of incisive analysis of industrial society, that mechanical engineers were capable of providing.[17] Dean of the engineering school at Cornell, past president of the Society, and one of the most distinguished figures in the progressively oriented wing of ASME, Kimball seemed the ideal person constructively to combat the Technocrats.

16 Ralph E. Flanders to L.P. Alford and 13 other ASME members, 15 Dec. 1932; Flanders Papers
17 George A. Stetson to Ralph E. Flanders, 12 Dec. 1932; Flanders Papers

Entitled 'Social Effects of Mass Production,' and designed as a rebuttal to Scott's 'Energy Survey of North America,' his speech directly engaged some of the most challenging issues of industrial capitalism.

Kimball admitted that the costs of manufacturing increasingly separated the worker 'from the ownership of the tools of industry,' and thus created the need for protective measures on behalf of that class.[18] He also recognized that another effect of mechanization was the replacement of skilled labor by unskilled workers. While such a 'degradation' process was as old as history, it had become particularly acute in the past century, suggesting to Kimball that 'the more rapid the rate of progress the greater the degree of suffering.' On the positive side, a larger amount of unskilled labor found employment in new kinds of mechanized industry and those people not only benefitted economically, but socially as well. In that fashion, America had successfully absorbed vast immigrant populations. This 'blotting paper' characteristic of technical advance, however, had been thrown out of balance by the depression, and Kimball worried that a decline in the development of industries based on new technologies might lead to more serious and prolonged unemployment. To prevent that possibility, he urged a shorter working week, agreements between producers to balance production and consumption, and social legislation.

Beyond these already familiar suggestions, Kimball saw another possibility on the horizon. He expected a reduced pace and scale of technological change in the future. As if the law of diminishing returns applied to technology, he envisioned 'an era of more moderate sized plants, less automatic and consequently more flexible in character.' Candidly admitting many of the charges so often levelled at mechanization, Kimball imagined the evolution of a more humane scale of industrial production, a kind of technology that by its appropriateness would eliminate most of the undesirable social effects of machinery.

Few of his colleagues followed Kimball so far, and Stetson, when he published the speech in *Mechanical Engineering*, edited that last part out. To give up the idea that prosperity depended on more technology meant giving up the idea that engineers were crucial to the recovery process. Instead, at least at a rhetorical level, the Society's leaders took the position, as Ralph Flanders put it, that 'the engineer made this civilization' and that it was up to him to save it.[19] In his own speeches around the country as ASME president,

18 Ms copy of 'Social Effects of Mass Production'; Flanders Papers. It was published in *Mechanical Engineering* (Feb. 1933) 83–9.
19 Ralph E. Flanders to G.W. Boston, 24 Jan. 1935; Flanders Papers

Flanders emphasized the necessity of rejecting radical solutions, whether they had to do with technical pace or political action. 'There is only one direction in which we can travel,' he told an audience of 400 Ohio engineers and steel executives at a meeting of the Youngstown Section, 'and that is forward, to greater economic, engineering, and financial development to obtain the greatest good for the largest number of people.'[20]

20 E.M. Richards to Ralph E. Flanders, 18 Jan. 1935, with attached newspaper clipping; Flanders Papers

In the iconography of mechanical engineering, the image of power and control is a popular one.

Ralph Flanders (1880–1970), architect of ASME's defense of Machine Civilization

TOP Clarence Ebenezer Davies (1891–1976)
Dynamic, outspoken, and passionately committed to the Society,
Colonel Davies served as ASME's chief administrative officer from 1934 to 1957.

Architect's rendering of the United Engineering Center,
on United Nations Plaza, in New York, a location that symbolized the
profession's internationalism just as the building's design was to reflect the
modern spirit of American engineering.

Whether in war or peace, ASME's leaders saw in industrial productivity
a central role for the Society.

LEFT, BOTTOM William L. Batt, shown here with a volunteer worker in the
CARE program, exemplified that attitude in his work for the War Production Board
and in foreign aid efforts after the war.

LEFT, TOP Dwight D. Eisenhower spoke for many of the virtues engineers cherished
and ASME responded by making him an Honorary Member, the first time
that distinction was conferred on an American president. Joseph W. Barker,
the Society's president in 1956, is shown making the presentation.

7

The Parker Case and
Institutional Confidence

When the members of ASME wrote or spoke about the depression, they often did so in the language of warfare. The crisis demanded that kind of national mobilization and suggested the curtailment of normal attitudes and practices for the duration of the conflict. Because mechanical engineers felt themselves under siege, notions of defense and counter-attack also came easily to mind. But there was a set of medical metaphors that proved equally attractive and appeared even more appropriate. In a way that implied ancient ideas of bodily humors, the depression was seen as a matter of imbalance; constituent parts of the system were out of harmony. And since mechanical engineers claimed credit for the economy's previous health, they saw themselves as the most likely physicians to minister to its ills. Furthermore, treating a sick economy, as the editor of *Iron Age* pointed out, called for an understanding of the true causes of disease, for remedies rather than palliatives.[1]

Yet engineers were themselves victims of the depression, and the gravest threat to ASME's own health was not from Technocracy or even from the widespread belief that mechanization caused unemployment. Loss of members and income posed more serious problems for the Society. A third of its members were without work and to stave off immediate ruin many were pressed to borrow $10 or $20 from a small relief fund the organization had started. But Council also insisted on a dues increase in 1930, when many members found their own incomes falling, and then balanced the budget by reducing services. Furthermore, even in those difficult days, the Society's annual meetings still had a certain style to them. Dean A.A. Potter's daughter remembered the banquet at her father's 1933 presidential inaugu-

---

1 'Engineers Plan Study to Determine What's Wrong With Industrial System' *Iron Age* (12 May 1932) 1053

ration as a glittering affair attended by the élite of America's mechanical engineering profession. These contrasts exacerbated the sense that the New York headquarters prospered at the expense of more distant members and that feeling took on the air of reality when the Society was sued for mismanagement by John C. Parker, a Philadelphia engineer and businessman with a taste for litigation. Initiated at one of the most discouraging moments in the organization's history, the action brought a court-directed investigation into ASME's business affairs and with its appeals and counter-suits lasted throughout the 1930s. As Clarence F. Davies put it, after the long, exasperating episode was over, 'the Parker case may be worth a chapter in a future Society History.'[2]

All of John Clinton Parker's business interests seemed to bring him into conflict with ASME's leaders. In 1914, when the boiler code was first advanced, Parker was engaged in the manufacture of steam-boilers and bitterly resented state regulatory measures as well as the Society's efforts to create standards for boiler makers. As he wrote the Council in that year, 'The writer desires to register a strong protest against further backing of the propaganda for state control of boiler design, with the funds, and at the meetings and in the publications of this Society.' Parker saw the work of the boiler code committee as the devious work of special interests to sabotage their competition and he darkly hinted at evidence demonstrating 'that these "interests" do not stop at underhand means to accomplish their ends.'[3] In a way that would become all too familiar, Parker's attacks on the boiler code were full of personal slander and extravagant language. And despite the committee's efforts to reach an understanding with him, it was never possible to achieve one. On the contrary, it was a mark of his behavior that, after having so vehemently opposed the code, he put an unauthorized boiler code stamp on his boilers and then charged commercial discrimination when the committee informed state authorities of the violation.

As it turned out, the boiler code disputes were just the beginning. Parker also published a kind of mechanical engineering handbook that brought him into frequent conflict with the Society. Called 'Lefax' and issued in the form of loose-leaf pages to be organized in a notebook according to one's interests, the venture proved highly profitable. But it depended mainly on ASME members, who were his principal customers, and Parker felt he ought to

2 Clarence E. Davies 'Comments on Eugene Ferguson's "Sense of the Past"' 3; ASME Archives
3 'Memorandum on the Parker Case' Harvey N. Davis to Officers of ASME Local Sections, 1 Sept. 1938; Council Minutes 24 Sept. 1938

have a special relation to the organization and the use of its facilities. Thus, for instance, in a 1926 circular letter to the membership he complained of the charge Calvin Rice made for the use of the Society's addressing machine: 'It is an astounding proposition to put up to a member who has been *scouting for data* for the membership for fifteen years.'[4] But Parker also used these broadside messages to members to advertise Lefax, while implying to manufacturing firms that the distribution of his data sheets was carried on under the aegis of the Society. Even more annoying to ASME's establishment, Parker argued in his letters to the membership that representatives of the rank and file, '*independent of any connection with the governing body or management*,' should pass judgment on the acts of those two groups.[5]

By 1929, this sort of troublesome activity had reached the point that the Executive Committee of Council empowered the secretary to inform all members 'that Mr. Parker has absolutely no connection with the publications of the Society, nor is he authorized to circularize Society members.'[6] But the fact is many of ASME's members used their connection to the Society for commercial advantage and that included the employment of membership lists and addressing facilities. Parker was different only in the extent to which he claimed a Society basis for his business activity and in the idiosyncratic nature of his approach to the organization's leadership. Engineers like to think of themselves as individualists, tolerant of differences, and the affair might well have ended, as one of his confrontations did, with an apology and its good-humored acceptance. But in 1933, Parker launched another attack on ASME that centered on its new secretary, Clarence E. Davies, and an alleged one-quarter-million dollar mismanagement of the Society's finances. At a time when the depression had already exacerbated the tensions between New York headquarters and more distant members, Parker's charges and attendant lawsuits raised serious questions for the organization and illuminated its state during those difficult years.

The long and expensive legal battle with John C. Parker had to do with the Society's publication of *Engineering Index*, an annual guide to technical periodical literature. ASME had purchased the *Index* in 1918 to consolidate its interest in processing current technical information and the step reflected the feeling within the organization that mechanical engineering was at the leading edge of America's great post-war industrial future. Through the next

4 Ibid
5 Ibid
6 Ibid

decade, the Society issued it regularly in yearly volumes. But in 1928, in another optimistic mood, the organization's leaders decided to expand *Engineering Index* in order to provide more extensive coverage of the literature and to make it immediately available through a current card-index which subscribers would receive in addition to the annual volume, for an extra charge. Although he never expressed himself clearly, the chances are that Parker saw a threat to his own business from the Society's publication of abstracts of current literature in a convenient form. That proposed activity was not exactly the same as Lefax, but Parker certainly saw himself as the principal purveyor to the mechanical engineering community of similar kinds of convenient, packaged information, and even before 1933 he had been criticizing the *Index* and the Society's publication activities.

The depression gave Parker the opportunity for a more direct attack. In its new form, *Engineering Index* lost money from the outset. Not only was it impossible to repay the Society for money loaned to begin the venture, but additional funds also had to be pumped in to keep it going. But in 1933, when ASME's own revenues had declined by more than one-half, and hopes for an economic upturn had been dashed, the Council decided to discontinue direct support of the *Index* and a campaign was started to raise outside funds for it. Andrey A. Potter, dean of engineering at Purdue and president of the Society that year, explained the budgetary problems in a circular to the membership, and his description of the accounts led Parker to send a letter of his own to members attacking the Council's stewardship. He charged that ASME's leaders 'had been sadly misled; that the Index had been put over with false pretences; that the funds of the Society had been wrongfully misused.'[7] Under such direct and unrestrained assault, the Council ordered the Committee on Professional Conduct to review Parker's communications to the members, obviously with an eye to deciding his own fitness for membership. The committee recommended no action against Parker, but at its meeting of 4 December 1933, the Council nonetheless expelled him from the Society, along with another of its critics, former Assistant Secretary William E. Bullock.

There is nothing to suggest any active collaboration between the two dissidents. Yet they shared opinions about certain issues that went well beyond the question of funding *Engineering Index*. Bullock, a naturalized American from England who had been assistant secretary for eleven years, strongly opposed Calvin Rice's ambition to generate enough advertising revenues to support the Society. But where Parker may have had some selfish motives

7 Ibid

for opposing commercial activity by the Society, Bullock's stand derived from ideas about professionalism. A vice chairman of the Conference of Association Executives, secretary for five years of the American Institute of Weights and Measures, and an active member of the AAAS, he argued that commercialism was inimical to a true professional status. In a 1931 pamphlet he sent to the membership, entitled 'Unofficial Proposals for Intensifying the Standing of Members of the Engineering Profession Through the American Society of Mechanical Engineers,' Bullock opposed advertising in Society publications and the publication of *Engineering Index*; he also suggested that all standards work be turned over to the American Standards Association, and that any ASME research which benefitted special interests should be financed by them.

Although Bullock stressed the importance of carrying out reform measures through the organization's normal channels, like Parker he in fact continued a direct mail campaign to persuade others to support his position. Later in 1931, for instance, Bullock wrote *Mechanical Engineering*'s advertisers urging them to cancel their contracts with the magazine. And in 1932 he sent out pamphlets to the membership with such titles as 'Is the ASME in the Hands of the Banks?' and 'Contortions in the ASME Finances.' The following year, he sent another circular letter asking members to pay only five dollars of their yearly fees, pending a constitutional amendment to lower the dues, which he planned to offer at the annual meeting.

Although made in the interest of an elevated style of professionalism, Bullock's attack on ASME's revenue-producing activities could hardly have come at a more delicate time. During the 1920s Rice had succeeded in raising 'outside' income to a level equal to membership dues, allowing for such expanded activities as the *Engineering Index* and a larger headquarters staff. But in 1933, with gross receipts reduced by one-half from the previous year, the Society's leaders were already concerned to arrest the decline in advertising sales. In fact, one of the earliest effects of the depression was to make membership dues once again the primary source of the hard pressed organization's income. Bullock's suggestion that only one-quarter of the dues be paid, and his threat of a constitutional amendment, thus presented an even more crucial problem than his attack on advertising, and the Council went to considerable effort prior to the 1933 winter annual meeting to defend their policies and marshal proxies to their side.

Such harsh dissidence has been rare in the Society's history. After all, except for instances in which younger men were pushed by their bosses into joining, members have been self-selected – they wanted to belong. Most of those who came into the organization and then subsequently found themselves out of sympathy with its operation simply dropped their membership. And perhaps even more to the point, those likely to become interested in ASME politics were men who already shared the attitudes and concerns of the leadership. Thus, in one sense, Bullock and Parker were isolated figures, unrepresentative of the Society's normal political life. Furthermore, open disagreement by staff members has been extremely rare and in Parker's case his behavior was at best eccentric.

Yet these episodes are historically important because dissent helps to reveal issues that are masked in the official record and gives tongue to rank and file members who are normally silent. Few people expressed themselves more plainly about the effect of Parker's lawsuit, for example, than George F. Holmquist, a loyal member from Bristol, Connecticut, who in 1934 wrote Ralph Flanders, 'From my point of view, the societye's business matters seem to be in a snarl. This man Parker from Phila. Pa. don't seem to help matters at all. For my part I would like to see somebody get in there and try to make it run smooth as they can.'[8] Holmquist expressed the desire of most members for effective leadership, especially at a time of such economic crisis, and like most members of the organization, he did not expect to be troubled about its management. The Council had hoped to restore that kind of tranquility by expelling Bullock and Parker. In Bullock's case, he retired quietly and the remedy worked. But Parker's reaction was entirely different. The Council's decision to expel him brought a legal action in the New York Supreme Court for New York County, charging the Society's leaders with misappropriation of funds, as well as the use of funds for purposes other than those specified in the charter, and he also claimed that he was still legally a member of the organization. Just as Morris L. Cooke had agitated the question of whether the Society served the public as it should, Parker argued ASME did not serve its members as they deserved. And to Council's utter dismay, in September 1934 Parker obtained a court order calling for a supervised investigation of the Society's financial affairs, an inventory of its assets, and his restoration to membership.

Judge Edward J. McGoldrick's ruling struck directly at the power of ASME's Council. In a manner typical of such organizations, the Society had long been governed by an oligarchy which enjoyed considerable latitude in its authority, both constitutionally and by a process of gentlemen's understandings. But the court ruled that Parker's expulsion was illegal on the grounds that the Council could not objectively decide his fitness for mem-

8   George F. Holmquist to Ralph E. Flanders, 5 Sept. 1934; Flanders Papers, Syracuse University

bership since it was an interested party in the dispute. That left the way open for anyone to get away with anything, according to an exasperated Clarence Davies, so long as they slandered Council while doing it. McGoldrick's order that the Society provide by 13 November 1934 an accounting for the previous year also invaded a long privileged domain. The treasurer had always presented annual reports, of course, but the amount of detail was limited and it was often arranged to support Council policies. In practice, few members actually paid much attention to financial reports anyhow, especially as the Society's affairs became more complex, and that reinforced the general tendency within the Council to control information. The court-mandated investigation, however, gave Parker the legal right to a wide range of details of ASME's headquarters operations and the opportunity to raise embarrassing questions about its management.

Parker's motives are not easy to assess. His pamphlets to the membership were long and rambling in an almost stream-of-consciousness style, mixing Lefax advertisement, denunciations of ASME's leadership, and an array of depression cures that blended populist, technocratic, and fascistic ideas. The words he was fond of coining suggest his thinking. Lefax, for instance, was derived from combining the words leaf and facts. He called one of his pamphlet series *The Enginera*, a shorthand way of saying 'the era of the engineer.' Similarly, 'Unista,' made up of the words united and states, was Parker's name for a proposed new and 'completely organized state in which everybody is employed and prosperity reigns.'[9] Thus, besides his business interests, which he felt were threatened by ASME's 'commercial' activities, he had notions about the engineer's role in society and visions of a new political and economic system in which the engineer predominated.

But if his concerns were diffuse, there was no doubt at all about the focus of Parker's attack. His legal actions named several Council members and past officers of the Society, but most of all, Parker's fight was with Clarence E. Davies. It was only one of the misfortunes which Ralph Flanders faced as incoming president in 1934 that Calvin W. Rice, ASME's secretary for almost thirty years, died that fall. But Davies, who had been assistant secretary since 1921, moved easily and naturally into Rice's job. His was a strong personality and men who subsequently attained high places in the Society still remember how rough Davies could be with eager young members whose ideas were different than his. In fact, part of what made his elevation so obvious was that he had already become the dominant figure of the headquarters staff – and that was the way Parker treated him, too.

9 *The Enginera* 35 (Jan. 1935); Flanders Papers

The court decision, Parker argued in one of his pamphlets, finally gave ASME members the chance for a full investigation of 'Mr. Davies' Management and of all paid help under him.' Suggesting a conspiracy between the new secretary and the Society's legal counsel, a bold headline in that message to the membership asked, 'Are Davies and Aaron in Control?' Another headline question with obvious point to unemployed engineers was 'Why Should ASME Employees be More Highly Paid than Members?' In Parker's mind, or at least as he claimed to members, it was the Society's 'paid help' who misled the Council, and he likened the organization to a financially failing business whose 'officers have kept control and have continued to draw big salaries while the shareholders were losing money.'[10] They were the real owners of the Society, Parker argued in a broadside he sent to all members, and they were the ones who should be exercising control over it.

Parker's pamphlet warfare was only one element of the battle. Legal maneuvers by both sides were simultaneously under way. At first, the Society's leaders hoped a 'wearing down' process of delay might put Parker off and in early December 1934 Aaron filed an appeal against Judge McGoldrick's order. Parker countered immediately with a request that the Society be held in contempt of court, but this was refused. During the curious sort of 'phoney peace' which prevailed into January 1935, Parker publically agreed at the annual meeting to withdraw his suit – but later denied he had – and then travelled to Springfield, Vermont, to meet with Flanders who hoped to settle things with personal diplomacy. Even at that point, however, Parker's attorney was proceeding with another appeal for a contempt of court citation and the second time he succeeded. Judge McGoldrick found past presidents Paul Doty and Roy V. Wright and secretary Clarence Davies in contempt, and jointly fined them $250 unless the Society complied with his original order by 1 April 1935. That ruling, plus McGoldrick's additional threat of jail sentences, led the Council to the Appellate Court for a stay of McGoldrick's first ruling, but they succeeded only in having negated the portion which restored Parker's membership. At that stage, there was nothing else to do but prepare the material Parker had asked to see.

The Parker case forced ASME to deal with the issues he raised. The most important of them, despite Parker's claim, was not political control of the Society; few members had any interest in a revolution to install Parker as 'the Huey Long of the ASME.'[11] What troubled mechanical engineers was

10 Untitled 31-page pamphlet, dated Lincoln's Birthday, 1935; Flanders Papers
11 Thomas D. Perry to Ralph E. Flanders, 8 March 1935; Flanders Papers

that they didn't know what was going on, a concern which was heightened by the sense of a New York clique, and even more obliquely, by the feeling that headquarters staff enjoyed ample salaries at a time when many engineers found their earnings much reduced. But the fact was that the information problem ran in both directions. Council did not have any better idea what members thought of the Parker affair than members had about what headquarters was doing. Therefore, the first step Davies recommended was a letter to local section officers soliciting the reactions of their groups.

At about the same time, McGoldrick's adverse decisions, especially the contempt citation, and Parker's pamphlets started to produce their own effect so that New York soon began to hear from widely scattered individuals as well as from those in the Society's political hierarchy. Generally speaking, members of Council and chairmen of standing committees, those closer to the top of the pyramid, preferred a minimal response. There were exceptions; Vice President Harry Wescott was eager to sue Parker for libel. But past president Conrad Lauer thought his charges should not be dignified with an answer while Alfred Iddles, a manager in 1935 and subsequently a vice president, argued that 'Parker's activities can be squelched by ignoring them.' Even Purdue University's Dean A.A. Potter, who claimed that the members were 'greatly disturbed,' still felt a one-page statement would be sufficient.[12] However, the further down the hierarchy responses came from, the more they urged some kind of action. A considerable number of section chairmen, for instance, replied that the Society's economic situation was too perilous and the charges too serious for anything other than a substantial explanation, although a number of them also recognized the erratic nature of Parker's conduct.

The same pattern held true in geographic terms; the harshest criticisms came from members farthest away from New York.[13] S.M. Marshall wrote from San Francisco to say he had 'believed for a long time that the management of the Society was extravagant,' compared to other organizations of which he was a member. He hoped for the return of a more simple program and that 'the number of employees and their salaries be reduced.' Orwell Logan, across the bay in Berkeley, claimed it was hard to know the merits of Parker's case from so far away, a feeling many members shared. But he pointed out that if Parker had not sent out his own pamphlets to the mem-

12 'Parker Case. Summary of Replies from Council and Chairmen of Standing Committees, March 12, 1935'; Flanders Papers
13 The responses cited or quoted from are in 'Resumé of Responses from Members to President Flanders' Letter of April 5 Regarding Parker Case'; Minutes of the Executive Committee of Council 20 May 1935.

bership, no one would have known the Council had expelled him. Logan then went on candidly to outline some of his 'misgivings' about the way the Society was run. Was ASME research disproportionately directed toward 'the light and power group of utilities?' he asked. Was there a group in the New York office that considered itself above the membership's control? And why, he wondered, couldn't employees' salaries be published in the annual report? Leigh M. Griffith, from the Los Angeles section, expressed a set of similar sentiments. 'I join in the general conviction,' he wrote Flanders, 'that the Society has been exploited by a group in the New York office, apparently for their personal benefit and profit, and that the members have been deprived of any effective voice in its control.' Griffith also thought that the real cost of *Engineering Index* had been concealed by 'clever bookkeeping,' and that staff and salaries should be reduced. 'After all,' he noted, echoing a point that emerged as one of the central concerns of the Parker case, 'this is a membership organization which the members are supposed to rule.'

Not all of the dissent came from the west coast, by any means. But it proved generally true that more distant members had more complaints. Part of that difference can be ascribed to the fact that eastern members who attended annual meetings and personally heard Parker's tirades discounted his pamphlet attacks on the Society and its management. But it was also the case that more distant members felt themselves excluded from the range of benefits easterners enjoyed, and excluded, too, from the mainstream of Society politics. Thus, while their criticism focused on the appearance of a New York clique and a top-heavy central administration, they did not limit themselves to those concerns. To some members, for instance, the boiler code seemed another example of a headquarters-inspired effort 'all out of proportion to the members' interests.' Several complained that *Mechanical Engineering* carried far too many articles on economic and political matters, while others objected to the number of engineering professors on important Society committees. The comments Ralph Flanders received also revealed that some members did not view the profession as the meritocracy it was so often proclaimed to be. J.H.V. Finney of Denver pointed out that many engineers thought the Society was managed 'by men who are no longer engineers but who, as presidents and managers of large industrial concerns, actually exploit the engineer in their efforts to cut labor and engineering costs.' And his was not the only complaint about the seeming bias of ASME's employment service towards employers.

If he ever worried about such things – and there is some evidence he did not – Clarence Davies might have thought it an impossible job to please all the members of ASME. The geographic problem, for instance, was inescapable and always had been. It proved equally inevitable that while a given number of members argued the importance of social responsiveness, about the same number objected to anything in the Society's publications but technical information. And the tension between academic and 'practical' engineers that had existed from the beginning continued unchanged into the twentieth century. Besides, against those kinds of chronic complaints, Flanders and Davies received a good deal of support from members, and many wrote in the spirit captured by A.H. Eldredge, a Boston consulting engineer, when he said, 'I have always and still do feel grand to be a member of the ASME.' Yet the Society had never before been brought to trial by one of its members on a charge of mismanagement, or, worse still, been judged in contempt of court. McGoldrick's order calling for an investigation of ASME's affairs did not in fact constitute a guilty finding, but there were a great many members who reminded Flanders of the familiar saying, 'where there is so much smoke there must be a little fire.'

That sense of underlying difficulty, even among those who staunchly supported the Society, pervaded the letters Ralph Flanders received in his capacity as ASME president. For instance, J.L. Alden, a Chicago Western Electric engineer, felt that Parker's pamphlet attacks were not the product of 'sound rational judgment,' but nonetheless concluded 'I think he has pointed a finger at certain weaknesses in our Society's arrangements.' An Ohio member, E.E. Kendall, also claimed he had 'no sympathy for such a rambling, meaningless appeal as Parker's. On the other hand,' Kendall added, 'he makes some criticisms that there are good grounds for.' And Thomas E. Cushing of Philadelphia emphasized the point that the Parker case revealed 'an undercurrent of opinion, namely, that the Society management for a number of years has been a disappointment.'

What most of those who wrote Flanders meant by mismanagement, however, was not inefficiency but rather that headquarters seemed to have a different set of priorities than those of ordinary members. Rank and file engineers found local section meetings, or those of technical divisions, the most important element of their ASME membership. And apart from the sociability of such gatherings, they tended to go to them for technical information directly related to their own employment. Practically speaking, those essentially local activities tended to define the limits of their professional concerns. ASME's Council, however, was filled with men who were ambitious for a nationally prominent place in their profession. Some were motivated by business interest or hopes for career advancement, but the dynamics of the situation demanded of practically all of them at least lip-service to the most professional of the Society's activities. The idea of an expanded indexing service to make the world's engineering literature more readily available was a project the Council and staff brought forward, not one that sprang from grass roots sentiment, and it was not the membership that pledged so much of the Society's treasury to the enterprise. And there were a good many members, according to William L. Dudley, who thought the boiler code less of an asset than 'an institution,' something New York was more interested in maintaining than they were. Dudley, chairman of the Committee on Local Sections in 1935 and a westerner besides, was close to the grass roots discontent Parker's lawsuit stirred up. Yet the future ASME vice president was also an organization man so that he saw the problem from both perspectives and, better than anyone else who wrote Flanders, was able to translate the membership's general anger over lack of information and lack of power into a number of specific suggestions that had to do with the Society's operations.

The first thing Dudley pointed out was that he had been able to defuse most criticism of the Society by personal contact, since a great deal of it came simply from an 'astounding ignorance' of policies and procedures. He implied, however, that New York was unaware of member discontent on a number of issues that were also fundamental in nature. The section chairmen he heard from, for instance, wanted to know why the Society's income and expenses could not be presented in such a way that a section officer could easily answer such direct questions as 'Are ASME dues too high?' or 'Are ASME members getting value received for dues paid in?' A number of the other suggestions Dudley relayed to Flanders also pertained to local sections. Members wanted to know why sections should not have a larger share of ASME income, so that local activities could be expanded without calling for individual donations over and above the high cost of annual dues? They argued that sections should have money for membership development. 'Why should the New York Office spend thousands on membership with decreasing membership? Let the *members* get *members*.' Dudley also forwarded the proposition that the constitution should be revised to give local sections 'a real voice' in the Society's operation.

The relation of sections to technical divisions – a question of enduring vitality in ASME – was another subject that emerged from Dudley's analysis of membership views. To those at the local section level, it seemed logical that the Society's geographic and technical divisions should work 'as a unit.' But they had more in mind than the customary notion that local sections should serve as an outlet for the technical papers of professional divisions. Those papers ought to have the same status as ones given at the Society's major winter and summer annual meetings. Indeed, among many of Dudley's correspondents, the annual meeting was seen as 'a farce' which added nothing to a paper and they claimed that local section and technical division meetings were the most successful form of Society gatherings.

Finally, Dudley reported a set of reactions that came back to the matter of management, although again from a local section's point of view. The members he heard from wanted to shift administrative emphasis away from New York. They argued for a small Council and a limited staff, with 'less money spent in New York and more in the field.' That not only meant a more active role for field offices; those writing to Dudley wanted the power of initiative and referendum for local sections, to give members some real weight in Society affairs. Thus, in a way no previous crisis had done, the Parker case raised important questions about ASME's headquarters staff and its relation to the members. One of Dudley's correspondents caught that mood very nicely when he said it seemed right 'that the work of running the Society should be shouldered by many with benefit to the member rather than by the Secretary with glory to the name of the Society.'

The actual work of responding to the court-ordered review of ASME's business affairs, however, clearly demonstrated how much the organization depended on its administrative staff. Parker's attorney had a mandate and he exercised it. Throughout the spring, summer, and into the fall of 1935, Clarence Davies and other staff members were compelled to marshal detailed financial information and to testify at 19 different hearings. In a way then, the Parker case came down to a struggle between Davies and his staff and John C. Parker. That became evident as ASME's headquarters employees increasingly became the target of Parker's shrill attacks. 'Davies Salary $9,000,' one of his pamphlet headlines proclaimed, and its story went on to give Ernest Hartford's salary as over $6,000 a year and George Stetson's as the same, calling the amount spent for staff salaries a piece of fiscal recklessness of the same order as the *Engineering Index* débâcle. 'How much could they earn,' Parker asked, 'in open and fair competition with members of the Society?'[14] He was right about the amount of Davies' salary in 1935. But Parker did not mention that as part of a general reduction of staff salaries during the depression the secretary had taken a $4,000 cut, and he did not learn that Davies' pay was raised to $10,000 a year in 1937.

The court hearings drew to a close just as the 1935 annual meeting preparations were under way and both sides maneuvered for advantage in what seemed the decisive battle. Parker sent out another pamphlet to the mem-

14 'To Members of the American Society of Mechanical Engineers $215,730.36 Gone' undated 32-page pamphlet; Flanders Papers

bership in an attempt to gain proxy votes that he might take to the meeting, and he wrote Davies for a committee room at the convention to use as his campaign headquarters. The Society's leaders also gathered their forces for the meeting, but in addition brought in Karl T. Compton, president of Massachusetts Institute of Technology, and Frank B. Jewett, head of Bell Laboratories – two of the most prominent figures in America's science establishment – to testify at one of the court's last hearings.

In fact, the dénouement came on 29 November 1935, just before the annual meeting, when the court-appointed referee, Judge Philip J. Sinnott, presented his report. Sinnott exonerated the Society completely. He found ample justification for *Engineering Index* in ASME's constitution and ruled that the Council had not acted improperly in setting it up, as Parker had claimed, even though evidence revealed the experiment had cost over $200,000. In fact, the report pictured the Society in quite a favorable light, and took pains to deal with Parker's restricted notions of professionalism. *Engineering Index* was such an appropriate task, Sinnott argued, that if it had not already existed the Council would have to invent it. Sinnott also found the work of the boiler code committee 'of supreme importance,' both to engineering and to the public welfare, and scolded Parker for his 'wholly unjustified' attacks on it.[15] That tone in the report and some of its actual passages, according to Clarence Davies, were the result of the testimony of Compton and Jewett and he suggested to President Flanders that he might 'again express appreciation for their services to the Society.'[16]

Neither Compton's and Jewett's endorsement nor Sinnott's findings altered Parker's sense of outrage, further stung by an assessment of $1,500 to cover ASME's hearing costs. Together with thirteen other members of the Society he brought a suit for damages in December 1936, charging Davies, the administrators of *Engineering Index*, and past members of Council with gross neglect of duty. As Judge Sinnott had observed, Parker's very profitable operation of Lefax made the loss of money in *Engineering Index* seem criminal, but he was also caught up in a quixotic crusade to re-organize the Society's entire publications program and with it the dues structure and headquarters administration. Thus, the lawsuit was actually brought on behalf of ASME members and any damages recovered against those responsible for the *Index*'s losses would go into the Society's treasury.

Sinnott's unequivocal findings might have suggested that this additional legal action by Parker was yet another example of his eccentricity, of no real

15 'Referee's Report on Parker Case (Dated November 20, 1935)'; Flanders Papers
16 Clarence E. Davies to Ralph E. Flanders, 27 Dec. 1935; Flanders Papers

danger. But Clarence Davies, who had to suffer lengthy examination, and the eleven past members of Council also called upon to testify, soon discovered otherwise. Mr Justice William H. Black of the Supreme Court of New York, in which the case was heard, relentlessly pursued the Society's officers to discover what they actually knew of the financing of *Engineering Index*. He discovered, for instance, that Charles Schwab, who had been elected president in 1927 because of his great success in managing the affairs of US Steel, 'showed an almost inconceivable ignorance of everything connected with the affairs of the society.'[17] Schwab was not the only official to be unaware of the ways in which the *Index* tumbled into indebtedness, and while there was nothing in the trial to suggest that anyone profited from such lack of knowledge, in his decision Justice Black found the Society's leaders guilty of gross negligence.

Black's observations revealed the difference between the way the law might view a membership corporation and the way one actually worked. He noted, for example, that the Council had never asked the membership whether to go on with the *Index*, and one of his rulings was that a 'letter-ballot referendum' be conducted by mail to discover what members did think – not only about its value but also even about its format and style. Black's decision sprang from his conviction that in organizations such as ASME, corporate officials were bound to act on behalf of the membership, just as in a democracy political officials were elected to serve the public. But historically Council had never sought the opinion of rank and file members on matters of policy or its implementation, or thought it appropriate to do so. What made Black's opinion even less appropriate for ASME was that because of its size and the geographic area it covered, headquarters staff had long since taken over the sort of duties the Justice regarded as the responsibility of the Society's elected officials.

Besides the fact that the defendants might be compelled to pay back into the Society's treasury all that had been lost on *Engineering Index*, Black's decision struck at several important structural elements supporting the organization's programs. No officer could serve effectively if threatened by damage suits of the kind Parker had initiated. And ASME's leaders could not accept the court's judgments regarding the letter-ballot referendum – a matter of internal management – or the Society's rights to acquire or dispose of property, such as the *Index*, acquired in the normal course of carrying out its constitutional mission. It seemed to ASME's legal counsel that these were solid grounds for an appeal, but the Society was also driven to do so because

17 *New York Law Journal* (13 June 1938)

it would have been impossible to admit Black's conclusion that many of the organization's most prominent men had been guilty of flagrant 'inattention, carelessness, and heedlessness of duty.'[18]

The next stage for the Parker case was therefore the Appellate Division of the New York Supreme Court and, with funds from a legal defense campaign launched by past president William L. Batt, that is where ASME's attorneys took their brief. Parker, in the mean time, wrote President Harvey N. Davis, demanding Clarence Davies' suspension 'until the disappearance of large sums of money have been explained to the satisfaction of the memberhip and of the High Court.' Parker also sent a circular letter to the membership calling for a business meeting of the Society on 26 September 1938, 'to examine the evidence in the ASME case, and to decide the questions which have been ordered by the Supreme Court of New York State to be decided by letter-ballot.'[19] It was relatively easy to ignore those taunts; Council had no intention of letting Davies go and the constitution defined the procedures for calling a special meeting. Furthermore, the Council had no evidence of member discontent equal to that of three years earlier, although Morris L. Cooke, who remained a persistent critic, claimed Black's decision reminded him of the dissipation of ASME trust funds which he had uncovered years before and simply provided another example that the Society 'has been run pretty largely as a private concern.'[20]

Thus, unlike the 1935 contempt citation, which had generated so much unfavorable criticism, Black's indictment of the Society's management did not stir any significant reaction from the membership. Perhaps that was because his ruling seemed so inapplicable to the organization, or perhaps it was because the original furor had pushed headquarters into a more active courtship with local sections and the communications gap between the two was not so great in 1938. In any event, when the Appellate Division reversed the lower court's decision in its entirety, and awarded ASME $2,500 damages against Parker to cover its court costs, there somehow seemed less to celebrate. In a matter-of-fact way, President A.G. Christie announced the court's verdict in an open letter to the membership in January 1939.

Of course, there was yet another reason for such a low-key response; no one knew if that was the end of the matter. In March 1939 Parker appealed the appellate court's decision and Council started a suit to collect from him

18 Ibid
19 'Memorandum on the Parker Case' Council Minutes 24 Sept. 1938
20 Morris L. Cooke to Maurice Holland, 13 July 1938; Cooke Papers, FDR Library, Hyde Park

the damages awarded the Society. In fact, however, nothing much came of either action. Parker's request was denied and although Clarence Davies pursued him until his death in 1946, and then his estate afterwards, the Society never recovered from Parker anything like the $30,000 it had cost to fight him. Even in his retirement, Davies still recalled the events with particular vividness. The depression was 'a rugged time,' he said. 'We were short of cash and paid our rent with notes. We sold bonds to the members. Rice and I took 30% cuts in salaries.'[21] And while *Mechanical Engineering* was full of articles about the crisis, the magazine carried nothing about its effect on ASME staff. The membership never knew, for instance, that Edna Murrayes, the backbone of the publication department's advertising sales, had her salary reduced from $250 a month in 1929 to $180 in 1933, or that it was not until 1945 that she regained her 1929 salary level. It was during just those desperate times for both staff and Society, when membership and income were so threatened, that Davies was compelled to deal with the Parker case. He spent more time in court than any other Society official and, more than on any other, the responsibility for its defense fell on his shoulders. It cost him five years of his time at one of the most crucial moments in ASME's history, he said, and he never forgot it.

There is a tendency to mythologize episodes such as the Parker case. In the corporate memory it becomes a story in which ASME's motives are misunderstood, one that suggests the trouble caused when people fail to recognize the Society's true purposes. Actually, it was the conjunction during the 1930s of economic and technological crises that infused Parker's attacks with historical significance. But even in mythic form the affair echoes an important element in the Society's experience. Depression era critiques of Machine Civilization, particularly the charge that massive unemployment was technological in its origins, made it imperative for the Society to defend itself. And there is a parallel between its public efforts to prove that engineering had beneficial results and its attempts to convince the members that it acted in their best interests.

Engineers remote from headquarters and threatened with lay-offs or salary cut-backs found it easy to imagine that the Society was in the hands of a small and self-interested eastern clique, particularly when the budget allocated to local sections was reduced even though dues remained unchanged. And it was in vain that Flanders argued to them that the organization's rules

21 Davies 'Comments' 2

provided for broad membership participation or that proportionately the east was actually underrepresented in Council. The fact was the membership had only limited knowledge of the Society's operations and Parker's pamphlets filled a crucial information vacuum. He succeeded in connecting the discontent members felt towards New York with that which they felt as a result of the depression's problems. The combination fueled his campaign, in spite of all its bizarre qualities, to turn out the paid staff and give the Society back to the members.

Just as those closest to the top of ASME's hierarchy favored a minimum response to the membership about the Parker case, the Society's leaders regarded the question of technological unemployment as another information problem. Most of them opposed anything which might sound like an admission of guilt, although from the outset a number of the Society's prominent men were concerned to find explanations which still left intact the long cherished proposition that engineering advances acted fundamentally in the public interest. One of the most favored defenses, epitomized in the fiftieth anniversary celebrations, was to dramatize engineering accomplishments, and the Society frequently chose a non-engineer to carry that message. Clarence Davies persuaded Charles A. Beard to edit *Toward Civilization*, for instance, and the celebration's planners selected Robert A. Milliken to be the principal speaker at the great banquet in Washington.

The George Westinghouse commemoration in 1937 was another explicit attempt to defend engineering against its adversaries. Westinghouse was the ideal figure for such a public relations effort. His railroad air brake, speed control devices, and his switching and signaling systems not only contributed directly to public safety, but also to the development of the American transportation network. He was as well a pioneer of the new electrical age and the founder of one of its giant firms. Best of all, those technical triumphs that brought safety and comfort also generated employment and economic prosperity. His own labors thus provided the object lesson to be drawn on the ninetieth anniversary of his birth.

James R. Angell, the president of Yale, delivered the main after-dinner speech of the occasion. In a familiar way, Angell linked engineering to the advancement of civilization and sketched out the destructive consequences of 'a prolonged moratorium in the field of scientific and engineering invention,' which some critics called for. The economic adjustment that technical change periodically made necessary was essentially an ethical problem to Angell. 'We are morally bound,' he argued, 'to insure that just as all segments of the community benefit from such advances, they should equally be protected from any of technology's harmful consequences.'[22] In Angell's mind, then, the depression's harsh consequences might well stir a sense of moral outrage, but not any legitimate criticism of engineering.

For someone who had never known Westinghouse and could not appreciate the technical details of his inventive accomplishments, Angell actually made quite an effective speech. Yet its conservative tone and predictable logic contrast sharply with a witty, irreverent, and fast-paced Works Progress Administration play staged in New York at almost the same time as the Westinghouse commemoration. Entitled *Power*, the production was a consumer-oriented attack on the private electric utility industry which Westinghouse had done so much to create. And even though that business was a major employer of ASME members, the play cast engineers in distinctly minor roles. Essentially faceless and portrayed simply as paid staff, they were the means by which utility companies were able to reduce their operating costs by 41 per cent during the years from 1913 to 1926, while only reducing rates 15 per cent.[23] One conclusion which might be drawn from the popular success *Power* enjoyed in a number of American theatres in 1937 and 1938 was that ASME's public relations were not much more effective than its membership relations. Or, to put it somewhat differently, under the stress of criticism from within and without, the Society's leaders found it about equally difficult to be completely candid either with their members or with the general public.

There were, however, some real constraints on Council during those years. In the first place, the Society could do little to solve the depression's problems. Just as there had been no mechanism to co-ordinate the membership's combined talents during the Great War, so there was none during the Great Depression. And efforts to create federal agencies for that purpose, such as the Science Advisory Board, ended in failure. But even if there had been some way to link the organization with appropriate governmental activities, neither the membership nor the Council could have agreed upon a policy; there were too many engineers who believed that a technical society had no business meddling in politics. Concepts of professionalism also interfered with the information process. Members of Council concerned to project an image of dignified objectivity thought it natural to establish an intellectual distance in public matters and that included a prohibition on airing such dirty linen as the Parker case presented.

22   James R. Angell 'Achievements of Westinghouse as Factors in Our Modern Life'
     Mechanical Engineering (March 1937) 161
23   Pierre De Rohan *Federal Theatre Plays* (New York: Random House 1938) 46

There were limits to what could be done about member discontent, too. It was not easy to discover rank and file opinion, to begin with. Even at the height of the Parker case, when Council's stalling tactics brought down the contempt of court citation, less than 1 per cent of the membership wrote headquarters to communicate any reaction – whether in support or not. A larger number expressed themselves to section chairmen and other officers so that some leaders had a sense of the membership's temper. But there were differences of judgment about that and in the absence of clearer signs it was difficult for headquarters to frame an effective policy regarding information to members. Other factors also inhibited the easy flow of news from headquarters to the membership. The Society's important decisions had always been made by a small group of men and the natural tendency of oligarchies to control knowledge was accentuated by the defensive position into which ASME's leaders felt themselves pushed by Parker's lawsuits and membership criticism. Besides, the Society was administered by a staff bureaucracy of increasing power, and it had a position to protect, too.

8

Whither ASME?

As the 1939 annual meeting of the Society approached, Clarence Davies found it natural to sum up his judgment of ASME's situation. The war in Europe was sure to involve the United States somehow. Besides, 1939 marked his twentieth anniversary in the organization and he had just seen it through five years of the depression as its chief administrative officer. He felt some pride in having maintained so many of the Society's programs and yet kept the budget in balance, although he also saw the harsh effects of long financial drought. Those might have been reasons enough for his candid and confidential memorandum to Council, but on top of them, Davies spent practically all his waking hours concerned with ASME and he thought he knew what it ought to be doing. And just as at the beginning of the depression he had been inspired by Charles A. Beard's *Whither Mankind*, at the end of it he entitled his own observations 'Whither ASME?'

The most crucial problem Davies saw was that engineers lacked 'a clear concept of the fundamental requirements of a profession.' The schools did not inculcate any sense of 'the social and ethical responsibilities of the engineer,' while lack of unity among the country's major engineering societies posed an equally great stumbling block to professionalism.[1] Davies identified 'a common purpose' as engineering's most compelling need, and pointed out that Vannevar Bush, president of the Carnegie Institution, had argued along the same lines in a recent speech before the American Engineering Council.

Within the Society itself, Davies said, there were some particularly serious weaknesses, the most obvious being 'that we have lost or are in danger of losing our technical leadership in some of our traditional fields.' In his estimate, only one-third of the technical divisions had 'forward-looking pro-

1 ASME Council Minutes 8 Dec. 1939

grams with a real sense of direction,' but even in those cases there was little connection to research or to codes and standards work in their fields. Because of severely limited funds the technical divisions also found themselves unable to compete with small specialist societies, only one new research project had been started in the preceding five years, and few leaders in industry knew anything of ASME.

Clarence Davies never was a man to mince words and he assailed the local sections, too. Their programs for recent engineering graduates were 'totally inadequate,' and he claimed that fewer than one-tenth of one per cent of local section papers were based on original research or were worth publication. And not only was the quality bad, but so also was the level of participation. Out of 614 men on local section executive committees, Davies noted that only 8 per cent were involved in any activity at a national level. The members of sections were no better; attendance at meetings averaged from 10 to 40 per cent of the members. Local sections, furthermore, did nothing about their 'real work,' which he saw as secondary school career guidance and programs of continuing education for the graduate engineer.

Besides these problems, Davies argued that the Society had no way of knowing what members thought about its policies and practices, or why so many resigned or decided not to pay their dues. Furthermore, service to the membership was seriously threatened by lack of staff, who had also to work in a crowded, badly ventilated office for reduced salaries and uncertain pensions. The answer to many of these problems was more money, but any significant increase in dues income might only be achieved by a lowered admission standard, and he opposed that.

Davies concluded his memo with a set of recommendations designed to promote a heightened sense of professional responsibility as well as more effective programs. But the emphasis was on the need to encourage ASME's technical activities. Besides suggesting ways to make Council more aware of the problems in that sphere of action and to stimulate technical research in the Society, the essence of his recommendations was that Council create a new form for the Advisory Board on Technology in order 'to develop an aggressive program to advance the technical life of the Society.'[2] Davies expanded on that aspect of ASME's work in a follow-up memo to the Council in June 1940. Warfare analogies seemed appropriate and he described the technical divisions as the Society's 'shock troops in our attack on technical problems.' Then, much as a commander might, he reviewed those institutional soldiers. The Steam Power and Fuels divisions were in top form, Applied Mechanics was 'vigorous and young,' Hydraulics, although a small

2 Ibid

group, was 'well led,' and the Railroad Division, another small unit, was also 'well officered.' In contrast to those effective divisions, Davies singled out the Machine Shop Practice Division and the Iron and Steel Division as examples of groups much in need of strengthening. Not only had those two divisions lost members to rival specialist societies, but they were also just the fields in which to attack 'the commonly held view that ASME is nothing but a power Society.'[3]

Those close to the Society's operations were rarely so direct in their evaluation. But apart from Davies' intention to survey the depression's damage, which he saw as having fallen mostly on the organization's technical activities, his memorandum was still clearly in the tradition Robert H. Thurston first elaborated. Just as Thurston had, Davies thought of the Society in terms of its two great essential elements – its field of social economy and its sphere of technical action. Thus, on one hand, he spoke of the social and ethical responsibilities of engineers, of the necessity for inter-society co-operation, and of better means for assessing and satisfying member interests – in other words, that mixture of concerns which included social status, political power, economic reward, and the management of the Society – which were usually expressed through ASME's geographic organization. On the other hand, Davies called for a vigorous program of technical activities across a broad front, ranging from original research and publication to the continuous absorption of new knowledge into the Society's codes and standards. That effort required the fostering of strong technical divisions, close enough to industry to recognize its problems but near enough to Council to obtain its support. In the same way Thurston had, Clarence Davies also thought of these two central objectives as closely related to each other. He gave the Society's social purpose a higher place in the scale of human values, but made its technical purpose the foundation of professionalism. It would have been impossible for him to imagine one without the other.

For many other of ASME's leading men the sense of standing at a historical watershed also encouraged that broad point of view. They still resented the way engineers had been so roundly condemned during the 1930s and, as war threatened a new world crisis, they seemed determined that the organization should never again be perceived simply as a technical society. George Stetson, editor of *Mechanical Engineering*, reflected that feeling when he argued the need for an 'intelligent and intensive concern with national and world affairs' to forestall any similar kind of criticism in the post-war era.[4]

3 Ibid 16 June 1940
4 'Engineers Face the Future' *Mechanical Engineering* (July 1941) 506

William L. Batt, ASME president in 1936 and head of SKF Industries, personified the wartime conjunction of the Society's social and technical purposes. In a 1940 speech that aimed to define the engineer's tasks in a 'world in flames,' he blasted 'business as usual' attitudes.[5] A Purdue graduate and scientific management exponent committed to ideas of social responsibility and a broad definition of engineering professionalism, he raised both the problem of accelerated wartime production and the equally important need for an orderly industrial demobilization in peace-time, to avoid a repetition of depression-era economic difficulties. Along with such other Society presidents as Ralph Flanders and Roy V. Wright, as well as staff members Clarence Davies and George Stetson, Batt urged ASME to acknowledge its economic and social duties to the membership and its professional responsibilities to humanity.

Some members of the Society argued that there still existed serious impediments in the way of establishing a vital social role for the organization. Gregory Dexter, a Scarsdale, New York, engineer, contrasted the hurdles he had to jump in order to get published a paper critical of a public utility with the self-interested article Lewis Sillcox, chairman of the railroad division and an executive officer of the New York Air Brake Company, wrote for *Mechanical Engineering*, arguing against canalization of the St Lawrence. 'He who criticizes the policies of the U.S. Government' Dexter claimed, 'appears to have no difficulty in being heard but he who dares to criticize a large public utility and important financial interests will have the greatest difficulty in being heard.'[6] Open access to knowledge was the crucial requirement for an effective hand in the nation's social and economic problems, according to Dexter, and he concluded his letter to *Mechanical Engineering*'s editor with another reference from Vannevar Bush's 1939 speech, previously published in the magazine, to the effect that engineering professionalism depended on its unprejudiced service to society. Although he felt himself an outside critic of ASME's ruling group, Dexter perceived, just as many of them did, that the coming war presented an important opportunity for a broadly conceived, public-spirited set of programs. And in common with them, too, he saw the war as an opportunity to extend a liberalizing process the depression had begun.

Besides its stimulating effect on the economy, the Second World War revived America's sense of technological mission and pride in its industrial

5 William L. Batt 'Through a Glass, Darkly' *Mechanical Engineering* (Jan. 1941) 5
6 *Mechanical Engineering* (Dec. 1941) 915

capability. In an important even if indirect way, that was one of the war's most telling effects on the American Society of Mechanical Engineers. Franklin D. Roosevelt's decision to set aside New Deal policies, to win the war by giving the great mass-production industries and their leaders essential control over the job of supplying the materials of war, was exactly the sort of judgment Society members could support, whatever they might previously have thought about the president's economic ideas. Donald Nelson, head of the War Production Board, put the matter plainly when he said the quickest way to get maximum output was to put men in charge 'who can understand and deal with industry's intricate structure and operation ... men with expert business and technical knowledge.'[7] That was a welcome message to mechanical engineers who firmly believed that crucial elements of productivity, such as standards or the organization of work, for example, had to be in the hands of those most familiar with the management of industry. Thus the war not only ratified a set of attitudes about expertise, but it also gave new power to the large industrial firms most mechanical engineers worked for.

This conjunction of forces – the movement within the Society to enlarge its field of social economy and the government's war production policies – seemed to be exemplified in the wartime careers of ASME activists. Batt, for example, went from the chairmanship of the Commerce Department's Business Advisory Council in 1941 to the Office of Production Management where he got the government's synthetic rubber program under way. When the OPM was dissolved, he moved over to its successor, the War Production Board, where he was finally responsible for America's total production of materials for the Allied cause. And true to his conviction that engineers had an equally important job in peacetime, Batt headed the Marshall Plan mission to the United Kingdom and served as the US member of the North Atlantic Defense Production Board.

Four other past presidents were in Washington, too. Among them, Harvey N. Davis headed the WPB's Office of Production Research and Development, while W.F. Durand, chairman of the National Advisory Committee for Aeronautics during the First World War and the man primarily responsible for the beginnings of aeronautical research in America, took over the chairmanship of NACA's jet propulsion committee at the age of eighty-two. And the careers of some of the Society's younger members similarly integrated technical expertise and broad responsibility. O.B. Schier, for instance, an assistant engineer with Consolidated Edison in New York and James N. Landis protégé who had been quite active in the Metropolitan

7 Bruce Catton *The War Lords of Washington* (New York: Harcourt Brace 1948) 117

Section, went to work for the War Production Board in 1941, helping to evaluate the wartime production abilities of small manufacturers. He did the same job for the Navy after he was commissioned in 1943, first in Chicago and then in Philadelphia. And just as Batt had done, when the war was over, Schier stayed on the job to assist in the sale and distribution of the Navy's surplus property. That kind of service was not only the logical way to employ someone with Schier's training, but an example for the Society's argument against the use of engineers in ordinary infantry assignments.

These wartime posts reinforced the conviction of ASME's leaders that the profession was destined to play a central part in the peacetime world. There was nothing new, of course, in that ambition to apply engineering training to a wide array of social and economic problems. It had found its first and clearest expression during the Progressive era at the beginning of the twentieth century, when engineers so actively connected ideas of professionalism and public service. After 1918, Europe's reconstruction had also inspired technical men to think that its special problems called for their special talents, and Herbert Hoover's post-war work in Europe gave them an example. Little more than a decade later, the Great Depression seemed to offer exactly the kind of opportunity that would allow engineers to demonstrate their ability to solve more than just technical problems. But what gave that old dream extra force in 1945 was the effectiveness with which the nation's scientific and technical expertise had been marshalled for the conflict – the atomic bomb was only the most dramatic example – and that seemed to insure a prominent post-war role for engineers in America.

In fact, the events of the next twenty-five years justified all those expectations. Continued high levels of defense industry research and development along with the simultaneous expansion of America's international interests, the spectacular growth of the economy, the Korean War, the space race, and the Viet Nam War – all tended to place a high premium on expert knowledge, to give the profession new levels of income and a secure sense of status. And in microcosm, one can see the tendencies of that quarter-century reflected in ASME's membership and income figures. By 1945, the number of members had climbed back up to 20,000, the same number that had belonged in 1930. During that period income had also returned to 1930 levels. In contrast, however, membership quadrupled over the next twenty-five years and by 1973 ASME earned ten times what it had in 1945, mostly from sources other than membership dues, another striking difference from the earlier period. Yet these remarkable changes in ASME's fortunes, paralleled by a set of alterations in the larger American society just as consequential, still unfolded against a backdrop of the organization's traditional historical elements. The Society's field of social economy and its technical sphere of action, those two central purposes that Thurston had identified at the beginning, have been as much a feature of ASME's modern history as they were of its classical era.

One reflection of the way these twin objectives remained entangled in the post-war era was in the ongoing struggle between geographic regions and technical divisions for political power. The depression years demonstrated the relative strength of the Society's geographic base. As hard times reduced travel to national meetings, local sections assumed more importance in the professional lives of individual members. And because ASME's leaders saw local sections as the primary source of membership dues, they were courted as never before. 'In these trying times,' Council assured local sections in 1933, their interests were 'always in the foreground of thought,' and it promised that even more attention would be given their point of view in the future.[8] When the Parker case threatened, it was again the local sections to which Council turned for support. Ralph Flanders made political necessity a virtue; one of the beneficial effects of the affair, he frequently observed, was that it put headquarters in better touch with its far-flung membership. But as the organization settled back into peacetime life, one of the first things Council did was to pay off that obligation. In a constitutional revision of 1946, eight regional vice presidents were created and given seats on Council. The change was justified on the grounds that it would 'shorten and simplify the relationship between members in the regions and the Council'; what it also did was to enshrine in the constitution an existing political reality, that the Society's geographic units commanded the largest share of budget and headquarters attention.[9]

The Society's leaders often deplored the political nature of the struggle between geographic and technical divisions, especially since, besides budget allocation, the competition also influenced the nominations process. In a long and thoughtful letter to the 1944 Nominating Committee, Clarence Davies felt compelled to warn it of the pitfalls in selection by pressure groups. Even if the basic principles by which it should operate were obvious, he reminded the committee that it ought to pick 'the best man for the job' and that 'the office should seek the man rather than that the man or his friends should seek the office.'[10]

8 Council Minutes 4 Dec. 1933
9 'Report to ASME Council of Special Committee on Society Organization Structure' *Mechanical Engineering* (Oct. 1944) 674–7
10 Minutes of the Executive Committee of Council 28 Nov. 1943

Whenever there was talk among engineers about unionization, a topic of increasing interest after 1945, one of ASME's spokesmen could be counted on to make the point that the organization was, after all, 'a technical society.' Yet the technical divisions themselves only slowly gained power in the Council, and it finally came not because Clarence Davies persuaded anyone of the need for a status equivalent to the geographic regions, but because of the growing importance of the divisions. In the years following the Second World War, younger men flooded into the profession and they pursued an ever widening range of special interests. This dramatic growth gave established technical divisions extra momentum and also created a demand for new divisions. Thus, just as in the decade or so after the First World War when Council was forced to respond to those centrifugal pressures or risk the loss of members to competing societies, so ASME's leaders in the 1940s and 1950s were pushed by new technical interest groups for recognition. The figures clearly outline the two comparable periods of growth. Twelve technical divisions were established in the decade or so after 1918, while nine were formed in about the same length of time after 1945.

Besides the direct lobbying of technical interest groups, the clearest signal to Council of these pressures for new divisions was what happened to junior memberships, the category intermediate between student and full member status. In 1951, for example, over 2,000 junior members dropped out of the Society, despite the fact that a new dues schedule reduced their annual fees, and a worried Council formed a Committee on Society Policy to review the organization's objectives and its structure. There was nothing especially novel in the creation of such a committee; a zeal for self-analysis has characterized ASME's twentieth-century history. What distinguished the 1951 committee, headed by past president Harold V. Coes, who had been active in the Committee on Professional Divisions, was its emphasis on the need for reform in the Society's technical sphere of action. In its report to Council, his committee argued the paramount importance of 'developing the specialized interests in the Society,' and it recommended that the problem be turned over to the standing Committee on Organization for the 'earliest possible' solution.[11]

Although the language was less than explicit, what the committee actually meant to propose was direct representation on Council for the technical divisions. It was a matter that stirred passions and they were mirrored in the Council's actions on the recommendations of the Committee on Organization. It had put forward a set of proposals at the winter annual meeting in

11 Council Minutes 30 Nov. 1952

1953, calling for an enlarged governing body to accommodate four positions directly representing technical divisions and two representing codes and standards. Just as important, the committee also proposed changes in the Committee on Nominations, since that group, comprised of men from the regions, was 'beholden to the geographic interests' of the Society.[12] And, in what might have been seen as a trade-off for technical division representation, the committee recommended that two additional Council seats be allotted to men who had shown leadership in the Society's 'administrative' committees, such as membership, professional status, education, or public affairs.

That last suggestion made the committee's report something everyone could disagree with. Men from the regions opposed the transfer of power to the technical divisions, whose men objected to representation on Council for member interest activities. As a result, and by a series of split decisions, the Council agreed to provide 'some degree' of representation for technical divisions, codes and standards, and the administrative committees; to refer the report back jointly to the Committee on Organization and the Committee on Constitution and By-Laws; and also by a split vote, expressed its confidence in the Committee on Organization.[13]

In the way standardization committees worked, the Committee on Organization took care to circulate the draft of its final report to all interested parties before submitting it to Council in the spring of 1954. And as with any standard, its conclusions were a compromise. The number of Council seats was reduced to its existing size and instead of six places being expressly allocated to the Society's technical sphere of action, it received four and the administrative committees also received four. From the perspective of the technical divisions, it might have seemed as if their concerns had been sacrificed to such things as public affairs and professional status, matters that tended to belong within the Society's field of social economy, while the number of regional vice presidencies remained unchanged.

Although engineers from the Society's technical sphere of action saw the battle for Council seats in political terms, to some extent it was also part of a post-war organizational problem that had become acute as ASME got bigger. By 1965, the membership had grown to twice its 1945 size, not counting student members, while the number of meetings, papers, and other technical activities had more than doubled. This extra work-load not only burdened the staff, but also raised questions about the Society's organization

12 Ibid 29 Nov. 1953
13 Ibid

and administration. The 1946 constitutional revisions had attempted to streamline things but, as the number of sections, regions, and technical divisions increased, the need for structural change became even more apparent. Besides, as 1960 president Walker Cisler pointed out, the Society was annually attracting only one-third of the country's mechanical engineering graduates and a substantial number of engineers dropped out after a few years' membership. In terms of ASME's administrative history, then, the 1960s were dominated by a concern for self-assessment and reorganization.

Cisler was one of several men from the Detroit Edison Company prominent in ASME affairs, and in common with many utility engineers he was also interested in management. His response to the problems of organization and membership was to hire a New York management consulting firm, Cresap, McCormick and Pagett, to analyze the Society's purposes and the way it implemented them. In essence, the firm advised ASME that it needed to 'clarify its objectives,' and recommended a series of changes to improve the Society's communications with its members, to integrate the work of staff and members, and to give Council better control over the various units of the organization.[14]

However, for an institution that always had a combination of objectives, the essential parts of which often ran in opposite directions, it was one thing to be advised that its purposes needed clarification but quite another to be able to do much about the matter. ASME had more than one goal because there were different interests and constituencies within the membership. Furthermore, it was not just the pull and tug between the social and technical purposes that confused things; the relation between staff and membership roles also remained an issue. Colonel Davies, as he was called after his wartime service in Army Ordnance, had a well developed sense of his own job and was never much troubled by that problem. But he retired in 1957 and by then the substantially altered scale of the Society's operations called for a more explicit definition of what staff was supposed to do. The major effort in 1966 to reorganize the Society therefore aimed to solve both the problem of technical division representation and the role of staff. But inevitably, as ASME also came to represent ever greater numbers of engineers who were corporate employees, finding a place on Council for member interests emerged as another organizational priority of the 1960s.

'Restructuring,' as the 1966 constitutional revision process came to be called, thus emerged from a mixture of necessities. Politically, the most important of them was action to redress the balance between geographical

14 William H. Byrne 'ASME: A Critical Look' *Mechanical Engineering* (Aug. 1961) 39

and technical representation on Council, although, characteristically, that issue was seldom acknowledged directly by the Society's leaders. Instead, Louis Rowley, chairman of the Committee on Organization's subcommittee that so carefully orchestrated the change, emphasized the need to define a more explicit role in ASME for its headquarters staff. Rowley drew a distinction between 'policy,' which he reserved for the membership, and 'operations,' which he made a staff function. Yet he did not mean to limit the staff role only to one of support service to the membership, any more than he meant to limit staff size. Members might still think of New York employees in those restricted terms, but Rowley argued for a greatly expanded headquarters operation, which at its senior level made 'the member role purely a policy-setting and advisory one.'[15] And just as Frederick W. Taylor had sixty years earlier, Rowley justified the increased cost of these extra administrative layers by claiming they would generate more revenue for the Society.

The key to his whole scheme was the creation of a series of policy boards, each with its own vice president, to administer ASME activities in the regions, technical divisions, codes and standards, education, communications, and professional affairs. That arrangement gave technical divisions representation on Council through the several departments into which they had earlier been collected and it also created a parallel set of staff directors who, Rowley suggested, would see the important jobs to be done and initiate action on them. Actually, the Society's senior administrators had already been doing that sort of thing for quite a while, but restructuring gave the practice formal status at the same time that it conferred greater political power on the technical divisions.

Restructuring aimed to create a symmetry between membership and staff and between the Society's technical and geographic interests. Each of those latter two groups, for instance, would form a sub-council of the main council, each was to have equal representation on the Executive Committee of Council, equal numbers on the National Nominating Committee, and each would also be represented on the staff by a deputy executive director. Rowley's plan was never fully implemented in that mirror-image form, but its parallel nature insured an efficient management structure and guaranteed that the tensions between different interest groups within the membership would be replicated in the staff organization.

Part of that drive for administrative and constitutional reform was thus a matter of old business. Technical divisions had certainly felt short-changed

15 'Restructuring ASME for a Dynamic Future' *Mechanical Engineering* (Jan. 1965) 23

for years and, in that sense, the 1966 restructuring rounded out a process begun in 1946 when regional vice presidents were created. But these revisions were also stimulated by America's sense of world power, on one hand, and on the other by the social and political turmoil of the 1960s in the United States. For instance, Walker Cisler saw the need for a review of ASME's management in terms of engineering's global responsibilities. 'Technology must be spread over the world,' he claimed, 'to correct, as much as possible, the imbalances that now exist in respect to population, natural resources, energy, and machines.'[16] The new United Engineering Center building, constructed in 1960 on United Nations Plaza, reflected the same spirit, and it was Clarence E. Davies, who had been particularly active in international engineering organizations, who selected the site.

Urban race riots, the anti-war movement, student demonstrations, and a swelling wave of consumer-oriented criticism of corporate industry proved an equally powerful set of forces that compelled mechanical engineers to rethink the purposes of their organization. As a politically and socially conservative group, they generally reacted strongly against the counter-culture style of protest in the 1960s, which seemed to flow from a set of values so antithetical to their own. Yet in a series of steps, the Society attempted to deal with the issues that its critics raised. In 1968, for instance, the Aviation and Space Division sponsored a session at the winter annual meeting entitled 'The Engineer's Responsibility to Society.' K.H. Hohenemser, a professor of aerospace engineering at Washington University in St Louis, identified the kinds of problems he thought engineers had a professional obligation to face. The road transportation system killed 50,000 people every year and injured another 2 million, he said, besides congesting the nation's cities and polluting its air. In the countryside, the mechanization of agricultural production left millions of less skilled workers without employment. A weapons industry created enough armament to destroy the world. But Hohenemser did not limit himself to the question of responsibility within the private sector – government had proved just as capable of disregarding the public interest. The AEC began an effective program of monitoring airborne radiation from atomic bomb tests only when forced to do so by public information groups. In a similar vein, he also pointed to the role of the Federal Aviation Authority in promoting a supersonic transport as more proof that neither government agencies nor industry could be expected to regulate themselves. And

16   Walker L. Cisler 'ASME Looks to the Future' *Mechanical Engineering* (Aug. 1960) 38

academic engineers concerned only with providing the kind of technical education industry dictated, Hohenemser said, needed to recognize their professional responsibilities, too.

In another paper of that program Adolph J. Ackerman, a consulting engineer from Madison, Wisconsin, argued that atomic power generation was being pushed too fast. He blamed both the utility industry and the federal government for creating a serious hazard and for doing so in a way that brushed aside the normal constraints in the private sector that acted to curb dangerous pursuits. The 1957 Price-Anderson Act, Ackerman asserted, relieved the utility companies of the financial risk involved in any atomic power accidents by transferring it 'to the victims and to the tax-payers of the nation.'[17] Without that legislation, insurance firms had refused to underwrite any portion of atomic power development, but, he claimed, as a result of it and of an aggressive effort by the federal government to promote peacetime uses of atomic energy, moderating influences within the utility industry and the engineering profession were suppressed.

Remarks of other speakers in the session, however, demonstrated the difficulty that confronted most engineers in dealing with the problem of their professional responsibilities. ASME members were largely industry's employees, not independent consultants as Ackerman was. If they went into managerial positions they could expect substantially greater salaries than those who remained in engineering jobs. But the higher up the corporate ladder they climbed, the more engineers were likely to experience conflicts between professional ethics and the business world. Besides, most of the Society's members had a strong sense of employee loyalty. J.R. Pierce, an engineer with Bell Laboratories, put that sentiment most explicitly: 'The engineer cannot allow purely personal interests or purposes, however noble they may be, to interfere with the function and purpose of the organization for which he works.' If a man did not approve of his firm's actions, Pierce stated, he ought to 'get out.'[18]

Pierce was not the only participant in the session to argue that ethics was an individual matter, leaving to the Society no larger role than that of framing general guide-lines. But that 1968 effort to define professional responsibility nevertheless became the basis for a standing committee of Council on technology and society which was ultimately elevated to the status of one of the Society's technical divisions.

17 *The Engineer's Responsibility to Society* (New York: ASME 1968) 37
18 Ibid 26

The general feeling of the 1960s that ASME needed in its organizational structure clearly to reflect a commitment to social responsibility found its most complete expression at the Goals Conference in January 1970 at Arden House, in Harriman, New York. By giving the conference such a title and by bringing to it the Society's most prominent leaders, the Committee on Organization meant to make an important change in the institution's direction.

The style of the conference came from a planning technique previously employed by, among others, the city of Dallas and by Richard G. Folsom, president of Rensselaer Polytechnic Institute. Perhaps as much to the point, Folsom deputized a member of his administrative staff, Rogers Finch, to assist the Society in carrying out the exercise. One of the central ideas behind that form of conference was to bring into the planning process all the constituencies touched by its results, an approach that was further heightened by the Society's upcoming centennial anniversary and the feeling that a broadly representative group would insure the future relevance of the goals they were to develop. On that basis, the organizers selected 90 people 'to represent ASME as a whole including its minority elements – not just its leadership.'[19]

Inevitably, of course, the conference was dominated by the Society's most well-known figures. Half those attending were or soon would be members of Council. Consequently, the group tended to be older rather than younger and almost entirely a gathering of white male engineers. Yet the spirit and conclusions of Arden House were far from conservative. As a result of the feeling that they had so often been outpaced by events during the 1960s, the conferees overwhelmingly argued that ASME had to be 'more activist, more dynamic, more outreaching.'[20] They expressed that sentiment in the form of an 'Over-riding Goal,' a great central idea of just the kind Donald Marlowe, ASME president and conference chairman, had hoped would emerge from such concentrated discussion. This goal which encompassed all other goals meant to describe the Society's principal future objective in unmistakable terms: 'To move vigorously from what is now essentially a technical society to a truly professional society, sensitive to the engineer's responsibility to the public and dedicated to a leadership role in making technology a true servant of man.'[21]

19 'Goals for a More Vital Society' *Mechanical Engineering* (April 1970) 17
20 Report of ASME Goals Conference and Proposed Statement of Goals, February 5, 1970;
     Goals Conference Papers, ASME Archives
21 Ibid

The language of the Over-riding Goal was such that Morris L. Cooke might have used; and, in a way also reminiscent of that earlier period of reform zeal, public responsibility was expressly allied to the economic welfare of engineers. Instead, however, of hoping for enhanced public appreciation of the profession – with tangible benefits – this new 'activist' Society imagined protecting its members against 'unprofessional employment standards and working conditions, against economic sanctions of various kinds, even against enforced idleness.' That uncharacteristically pointed way of speaking typified the Arden House Goals Conference. Those attending recognized that modern technology impinged more directly on more people than ever before, that the public was 'less tolerant' of such hazards than ever before, and that the individual engineer might well 'jeopardize his position, or even his career, by standing for what his engineering conscience tells him is right.' But the conferees agreed it was time for the profession to turn from 'its traditional caution' and they put forward two additional goals to buttress their position. One called for the Society to take stands on public issues that involved technical knowledge and the other would have it 'provide support' for engineers willing to put conscience ahead of their careers.

The passion for a Society more sensitive to its professional responsibilities, both to the public welfare and to that of its members, pervaded the goals set down for a number of other ASME activities. In the case of membership requirements, for instance, those at Arden House took the position that anyone 'engaged in mechanical engineering' should be eligible to join the organization. Such an approach not only recognized the growing number of technicians and engineering technology graduates who were not affiliating with ASME, but it also rejected the argument that a restricted membership enhanced the status of those who belonged. In a similar fashion, the conference agreed that the Society should investigate the use of new communications technologies to provide better service to members and renovate its publications so that they reflected 'this expansion of concerns beyond the purely technical.' The Society's relations with government were also touched by the same spirit, in a goal which called upon ASME 'To provide government at all levels with technical advice in the public interest and to develop a climate of understanding and credibility that will foster a continuing dialogue.'

ASME's 1970 Goals Conference had emerged from the conjunction of several concerns: George Habach, the Society's president in 1968–9 and a Studebaker-Worthington vice president, was a strong adherent of long-range planning; Louis Rowley, who had preceded Habach as president, argued that the continuity of programs demanded an enlarged and more active staff;

Donald Marlowe, Habach's successor, thought ASME needed a new sense of direction and dedication. More than any others, these three men shaped the Arden House proceedings and what flowed from the event was a substantial commitment to a broadly defined professionalism that implied a more public role for the organization, more active efforts on behalf of the professional and economic welfare of the membership, and more staff members to carry out those initiatives.

The reaction to the conference suggests that ASME's leaders had perceived a role for the organization which squared with the opinions of many members. Letters to the editor of *Mechanical Engineering* focused especially on the goal that related to member interests. L.J. Reggiardo, a California engineer who supported unionism, argued that the way 'to revitalize' the Society was by using its collective strength to increase salaries, rather than by 'motherhood' statements on the importance of the profession.[22] Others wrote to propose that responsibility for the membership's economic and professional needs be incorporated into the constitution. But a 1971 petition, signed by 101 members, to make it the Society's primary purpose 'to promote and improve the economic well-being of its members,' was decisively defeated at the summer annual meeting.[23]

To some degree, this emphasis on economic security came from engineers thrown out of work by cut-backs in military spending and the space program. It obviously did not represent the attitudes of those at the 1971 summer meeting and, besides, there were others who wrote in opposition. J.F. Sapp claimed that 'the proposed goals would establish ASME as a combination labor union, insurance organization.'[24] Yet a 1976 survey of the membership conducted by Georgetown University Poll revealed substantial sentiment in favor of increased attention to public policy matters and to such member interest issues as the portability of pension plans, salary improvement, liability insurance, and certification. Furthermore, an analysis of member response showed a striking difference in attitude between the 8 per cent who had been active in the Society for a number of years and the 80 per cent 'who have never worked on an ASME committee.'[25] Those with a long record of participation were strongly opposed to a public role for the Society, to activism in political affairs, and to the extension of member interest concerns. The great majority of the rank and file, however, warmly supported such initiatives.

22 *Mechanical Engineering* (Jan. 1971) 57
23 Ibid (April 1971) 67
24 Ibid (July 1970) 87
25 'ASME Member Survey Finds Out' ibid (Sept. 1976) 38

These differences of opinion over purpose and direction – whether technical action or social economy should predominate – were almost as old as the organization itself, and they were not limited to ASME. The recession conditions of the late 1960s, which had particularly affected engineering jobs dependent on federal spending, also led the American Institute of Electrical and Electronic Engineers to take up the cause of its membership's economic welfare, and a constitutional proposal to devote the institute largely to those kinds of activities was only narrowly defeated in 1971. But especially for ASME, whose history had been so much shaped by the interaction of social and technical purpose, a chain of events during the closing decade of its first century cast those two outstanding elements of the Society's life into sharp and contrasting relief.

The Goals Conference at Arden House had begun the process. Its prevailing theme had been the need for a change in emphasis and, of all the participants to speak to that issue, Professor Stephen J. Kline of Stanford University most explicitly compared the Society's two great spheres of action. In a mock report-card assessment, he graded ASME's performance across the board. He gave its efforts to advance knowledge high marks, particularly for the quality of its published papers and for its work in codes and standards. The Society's 'usefulness to the profession,' judged in four different categories, received failing grades, however, as did its educational programs, its attempts to develop ethical standards, and its activities that ministered to the public interest. Kline meant by his report card merely to draw attention to his claim that while the organization had done 'an excellent job on technical matters,' it had failed to protect its members against their corporate employers and failed to deal with the larger public problems modern technology had created.[26] Furthermore, as if the Progressive Era had never existed, Kline argued that ASME's local sections were the ideal vehicle for civic-minded engineers to use in applying their expertise to such matters as air pollution, urban transportation, and natural resource conservation.

The Goals Conference emphasis on member interest was so much more pronounced than anything that had gone before that it raised questions about a consequent threat to the Society's tax status as a non-profit, scientific and educational institution. The preponderant feeling of the conferees, however, was that the Society had been 'too cautious' about its tax status. They called that fear 'the 501 (c) (3) Syndrome,' and identified it as an undue concern with the tax-exempt, non-profit aspects of the classification, which had led

26 'Comment by S.J. Kline on "ASME: Professional or Technical Society? by Clarence E. Davies" '; Goals Conference Papers

engineering societies 'to disassociate professionalism from economic status' and prevented them from addressing 'controversial questions of public policy.'[27] Although the opinion did not surface in the final report from Arden House, some of the Society's leaders even argued that it would be better to have a 501(c)(6) – 'Business League' – classification instead, so that ASME might more actively pursue its interests in the political arena.

But when in 1977 the Internal Revenue Service moved to revoke the Society's tax-exempt status – along with that of the American Society of Civil Engineers, the American Institute of Chemical Engineers, and several other scientific and technical organizations – ASME's Council decided to fight the ruling. By that time, the Society was confronted by a serious lawsuit involving the boiler code that provided embarrassing ammunition for the critics of its standards, and because of its heavy commitment to nuclear power, was also on the defensive from attacks on that industry.

Those closest to the work of the boiler code, ASME's oldest and most prestigious safety standard, were always acutely aware that it might be compromised by the undue influence of interested parties. It was not a question of eliminating special interests – their suppression was even undesirable in the context of a voluntary standards system – but rather of effectively balancing them. And as the institutional structure that generated standards in America grew more complex, the techniques to assure an orderly evolution of a given standard became better defined. But legal challenges to the objectivity of standards periodically forced the Society to examine its own procedures. The preliminary report of a 1963 subcommittee to review codes and standards activities recommended, for instance, that the Society document its efforts to secure adequate public interest representation on codes and standards committees, as a defense against the charge of negligence, that it take care to protect the interests of small producers and consumers, and that it carry liability insurance to protect itself from any damage claim.

Yet against the increasing threat of litigation, ASME's leaders could take some comfort from a recent court decision involving the Johns-Manville Corporation and, as an unindicted co-conspirator, the American Society for Testing Materials. In acquitting the firm of the charge of restraint of trade, the court also found that ASTM's procedures made it 'likely that the results reached by them will be scientifically sound and will represent the general interest.' Even more appealing, the court stated that because government at all levels relied so heavily on ASTM for technical specifications, it might

27 Report of ASME Goals Conference and Proposed Statements of Goals; Goals Conference Papers

reasonably be regarded as 'an essential arm, or branch of government and its acts may be entitled to the immunity from the anti-trust laws accorded government acts.'[28] Therefore, the court concluded, rather than assisting the proper formulation of standards, the threat of anti-trust prosecution actually interfered with the work of the society's members.

No decision could have been more heartening to ASME. The subcommittee reviewing its codes and standards work claimed 'if these facts are true for ASTM they should be even more applicable for ASME, because of our broader areas of interest.' And to the Society's legal counsel, the dangers of anti-trust prosecution also seemed remote if committees rigorously followed the same American Standards Association regulations that governed ASTM and most other standards-producing bodies.

In 1975, however, the Hydrolevel Corporation, a small manufacturer of boiler feed-water indicating devices, brought just the sort of lawsuit the Society had hoped to avoid, suggesting precisely the kind of collusion that critics of industry-developed standards claimed was inevitable.

The essence of Hydrolevel's charge was that in 1972 McDonnell & Miller, Inc., an IT&T subsidiary that controlled 85 per cent of the market for heating-boiler safety controls in the United States, conspired with a subcommittee of the heating-boiler code committee to secure a ruling adverse to the sales of its product. In particular, the case hung on the fact that the chairman of the subcommittee on heating boilers, T.R. Hardin, a vice president of the Hartford Steam Boiler Inspection and Insurance Company, together with the vice president for research of McDonnell & Miller, J.W. James, who was also the vice chairman of the subcommittee, jointly drafted a letter from McDonnell & Miller to the subcommittee requesting an interpretation of that section of the code which dealt with devices of the sort Hydrolevel and McDonnell & Miller manufactured. The subcommittee's response was written by its chairman, Hardin, answering himself, as it were, and Hydrolevel claimed that his letter was subsequently employed by McDonnell & Miller salesmen to discredit the competing device. But the matter did not end there. Hydrolevel complained to ASME about those sales tactics, only to have McDonnell & Miller's James, who by then had become chairman of the subcommittee on heating boilers, aid in drafting another response to Hydrolevel.

In its own effort to resolve the problem, ASME's professional practice committee operated without the crucial information that subcommittee person-

28 Supplemental Report to Organizational Committee of ASME, October 8, 1964; Folsom Papers, Rensselaer Polytechnic Institute, Troy, NY

nel had drafted the original letter of inquiry as well as the responses to it. And from the Society's point of view, the situation was further complicated by a *Wall Street Journal* article in 1974 that condemned both ASME and McDonnell & Miller. From that stage, the affair went to a Senate hearing before the subcommittee on anti-trust and monopoly and then to the courts. The other two defendants in the case, International Telephone & Telegraph Corporation and Hartford Steam Boiler Inspection and Insurance Corporation, settled out of court in 1978 for a total of $800,000, of which IT&T paid $725,000. But ASME, which had continued to fight the matter, was assessed damages of $3.3 million, trebled by federal law to $9.9 million, when a United States District Court ruled in 1979 that the Society was guilty of violating anti-trust statutes.[29]

Both the enormity of damages awarded and the strong appearance of conspiracy came as a profound shock to ASME's leaders and its members. Just as in the Parker case, almost fifty years earlier, the Society appealed the decision and it tried to reassure its members with a special report in *Mechanical Engineering* in April 1979 from the president, O.L. Lewis, bravely entitled 'Bloody but Unbowed.' Like the Parker case, too, although there were also obvious differences, Hydrolevel's suit once again raised questions about the control of knowledge – whether pertaining to the way ASME conducted its business or to its role as a steward of technical information. Melvin R. Green, managing director of the Society's codes and standards operation, complained that 'the public doesn't understand that our principal purposes for developing codes and standards are to protect public health, safety, and welfare.'[30] And because of that misunderstanding, he said, government would continue its efforts to regulate standards activities in the United States. But it was not only the general public that the Hydrolevel case disturbed. Angry members wrote to the editor of *Mechanical Engineering* to protest against the Society's lack of candor in dealing with the affair and against codes and standards procedures. As one of them put it bluntly, 'Why was John W. James appointed or elected to various offices and chairmanship of the ASME B & PV Code Subcommittee on Heating Boilers when he was a manufacturer of heating boilers? This is like appointing Babcock & Wilcox, GE, Westinghouse, Combustion Engineering and General Atomics to the Nuclear Regulatory Commission. The fox guarding the chickens.'[31]

29 For details see Nancy Rueth 'Ethics and the Boiler Code' *Mechanical Engineering* (June 1975) 34–6, and *Wall Street Journal* 9 July 1974 and 7 April 1975.
30 Melvin R. Green 'Evolution of ASME Voluntary Standards and Their Future' *Mechanical Engineering* (June 1979) 108
31 Howard Moore to J.J. Jaklitsch, 14 May 1979; ASME Archives

In fact, however, the intricate web of relations between technical organizations, industry, and government in America has often been characterized by just such apparent contradiction. The majority of engineers who have the specialized knowledge and experience to frame effective standards work for industry. And it is exactly that connection to business and to its interests which insures that those most familiar with a product, both in terms of technical requirements and market-place considerations, are the ones who define its standards. ASME's role was to balance the relevant interests in a context which advanced both the art and the public's welfare. Thus, just as profit provides a mechanism for the allocation of the factors of production, the partnership of industry and technical society in drafting standards was, ideally, to create a system that harmonized the forces of technical change and the need for technical order.

The Hydrolevel case was hardly an example of the system at its best. To paraphrase an old legal maxim, standards had not only to be impartial, they had to be seen to be so. Instead, it appeared that large firms with large engineering staffs used their dominant position in ASME's codes and standards activities to drive their competitors out of the market. And it was the appearance of things which stimulated a growing movement among public interest groups for federal regulation. Mel Green, who spent an increasing amount of his time in the 1970s defending the voluntary standards systems at congressional hearings, claimed that the regular, American National Standards Institute–sanctioned procedures (after it replaced the American Standards Association as the principal umbrella organization in standards) had been followed in the Hydrolevel case. But even as the Society's leaders waited for the verdict on their appeal of the District Court's judgment, the fundamental issue in Green's mind was still whether the country was better served by government-imposed standards – a prospect that conjured up endless bureaucracies staffed by the technically inept – or a system of voluntary standards which in a 'more democratic and a more cost-effective' way protected the public interest.[32]

ASME's credibility as a steward of technical knowledge was even more sharply challenged by critics of nuclear power. In a letter to the *New York Times*, for instance, consumer activist Ralph Nader declared, 'to claim that the ASME Nuclear Committee is an objective and balanced body is patently absurd.' B.F. Langer, of Westinghouse's nuclear division, and a member of

32 Green 'Evolution of ASME Voluntary Standards' 108

the Policy Board, Codes and Standards, argued in rejoinder that 'anybody without some experience in the nuclear industry would not be competent to serve on this committee.' And Langer pointed out, once again, the central thesis behind industry developed standards: if those with any conflict of interest were excluded from the process, all engineers 'with any deep knowledge of the subject' would also be eliminated.[33] Furthermore, Langer said, the Nuclear Committee did not draft standards – that was done by appropriate codes and standards committees, and they were written in strict accordance with American National Standards Institute rules.

Nonetheless, ASME was deeply involved in nuclear power. The generation of electricity had always been one of its major areas of interest and the Committee on Nuclear Energy Application, formed early in 1946, was chaired over the years by such prominent Society figures as Alex D. Bailey of Chicago Edison, Walker Cisler of Detroit Edison, and James N. Landis of the Bechtel Corporation – all of whom were ASME presidents at one time or another. Nuclear power also heavily involved such long-time supporters of the Society's codes and standards work as Babcock and Wilcox, General Electric, and Westinghouse. And if all those connections did not guarantee a substantial commitment to the new energy source, the sale of nuclear codes and standards did. By 1979 the Society's annual income from codes and standards publications was well over $10 million, most of it from the nuclear power industry.

But the challenge to the Society's relations with the industry did not come only from anti-establishment critics. Within the organization itself there were differing views on the way ASME should handle information connected with the subject. In February 1976, for instance, *Mechanical Engineering* published two related items on nuclear energy. One was an article by Ralph Nader entitled 'Nuclear Power: More than a Technological Issue.' It was followed by a response from the nuclear codes and standards committee. Industry engineers, however, including Langer, raised strong objections to the publication of material unfavorable to nuclear power interests and they felt betrayed by the decision to give Nader space in the magazine. 'It may be the duty of the mass media to present both sides of controversial issues,' Langer argued, 'but it is the duty of ASME to present expert judgments and conclusions based on facts, not lies or fantasies.'[34] To prevent the recurrence

33 The exchange between Nader and Langer appeared in the *New York Times* 10 April, 29 April, and 4 May 1976.
34 Bernard F. Langer to Policy Board, Codes and Standards, 23 April 1976; Folsom Papers

of another article such as Nader's, Langer called for the appointment of an editorial board to make certain *Mechanical Engineering* reflected 'the considered opinion of the Society.'

J.J. Jaklitsch, the magazine's editor, defended his choice on the grounds that nuclear energy was a legitimate news story and that the aim of *Mechanical Engineering* was 'to make people think.'[35] But Langer, who won considerable esteem from his colleagues for his contributions to the development of nuclear codes and standards, typified a viewpoint which gained increasing currency in the 1970s. The Hydrolevel case, as well as critical attacks on the voluntary standards system, nuclear power, and indeed on a whole range of industrial activity that had once seemed a cause for national pride, gave strength to those within ASME who favored a more conservative approach to public issues than the goals conference had recommended. Dudley Ott, a Bechtel Power Corporation executive and ASME vice president in charge of the Power Department, spoke to that theme when he warned those on his committees responsible for reviewing technical papers to be careful of statements 'which could easily be misinterpreted by the press or environmental activist groups.'[36] In the same vein, the Policy Board, Codes and Standards, vetoed a 1975 proposal by the Public Affairs Committee for a 'free wheeling' panel at the winter annual meeting that would have included 'outside activists.'

The argument that Langer, Ott, and others like them used to support their position was that ASME was a 'technical' society. Just as the word 'voluntary' in connection with standards had political connotations, so in that other context did the word technical. What they meant was to limit the Society's field of social economy, and particularly in those kinds of activities that tended to loosen control over technical knowledge. Thus, the men from the Power Department wished for stricter guide-lines in the subject matter of technical papers, opposed public statements on power generation by the National Society of Professional Engineers or any other group devoted essentially to member interests, and felt that the Society's Washington office should not 'tell Congress what to do,' lest its tax status be jeopardized.[37] But they supported with ease the Society's official opposition to Proposition 15 when California voters were considering a ban on nuclear power development.

That was the dilemma for those who would curtail the Society's social purpose. The only basis on which they could acceptably claim the right to

35 J.J. Jaklitsch to Policy Board, Codes and Standards, 29 April 1976; Folsom Papers
36 Dudley Ott to F.W. Nelson, 5 Oct. 1973; from Mr Ott's private papers in his possession
37 Dudley Ott to Robert E. Roberts, 4 March 1975; Ott papers

control technical knowledge was that specialists alone could understand its implications, and that as professionals they acted in the public interest. But stewardship unavoidably called for a public role for the organization and it had to be seen to be an impartial one. Besides, there were members of the Society who favored more action in the field of public policy, not less, and judging from the 1976 member survey, their numbers were not small.

ASME's centennial anniversary provided it with an obvious opportunity to address the problems it faced in 1980 and also to reassess the goals it had so positively defined a decade earlier. Just as its leaders had a half-century before, centennial organizers scheduled an event at the McGraw-Hill building in New York and another to follow at Stevens Institute of Technology across the river in Hoboken, to commemorate once again the historic steps that had led to the organization's founding. But there was nothing planned to dramatize the heroic triumphs of engineering, nothing to demonstrate its centrality to the human experience. Instead, in a move that better reflected ASME's modern style and temper, the Committee on Planning and Organization proposed a conference entitled 'Strategies and Structure for Century Two,' and engaged the Arthur D. Little Corporation to send some of its staff along in the event the services of a management consulting firm might be required.

Held in August 1978 at a lakeside conference center in Ohio, the 'Century Two Convocation' resembled the Arden House gathering of 1970 in a number of ways. Edward H. Walton, ASME's deputy executive director, arranged both conferences and that had something to do with their similarity. The notion of assembling what purported to be a representative gathering of mechanical engineers for intensive discussion of the Society's future owed much to the Goals Conference, too, which also left behind it a momentum that helped shape the agenda for the Ohio meeting. But the problems that those who went to it discussed, the issues that stirred them to debate, had a timeless quality.

The proliferation of smaller technical societies suggested to the conferees that ASME had not yet found a way to control the centrifugal tendencies of engineering specialization. The threat of big government and the sensation that the Society was always a step behind major social and political currents rekindled the old hope that more creative kinds of responses might be possible. It still seemed to industry-based engineers that technical educators were out of touch with some important realities, just as the teachers argued that industry's failure to encourage professionalism left the way open to unionization. Those attending the Century Two Convocation worried about the vitality of the winter annual meetings, the lack of effective interaction between local sections and technical divisions, and the state of the Society's honors and awards program. The question of engineering unity was another familiar topic, as were the arguments advanced for and against it. The Society had yet to employ modern communications techniques to discover why, for instance, so many members dropped out after about seven years, and the profession still failed to attract significant numbers of women or members from minority groups.

Besides all those well-known issues, the ones that emerged as dominant themes were also common to ASME's history. A concern to devise better ways for bringing younger members into active participation, to decentralize the Society's operations, and to streamline its administration were the problems which seemed most in need of correction. But whereas the Goals Conference was inspired with a mood of great optimism, the Century Two Convocation seemed more characterized by the feeling that the times ahead would be harder. That sense of things was also reflected in the management study conducted by the Arthur D. Little Corporation, which had naturally flowed from the convocation judgment that the Society's organization badly needed overhauling. In its report, the management firm forecast even more litigation, regulation, and legislation. It predicted a closer scrutiny of tax-exempt organizations in the future and a growing number of public issues to entangle the Society.

All these changes demanded that ASME change, too, the report's authors argued, but their prescription was actually one of retrenchment. Without a sense of the historic nature of the Society's dual mission or an appreciation of the way the creative tension between social and technical purpose had vitalized the organization over the years, they concluded the former was a branch of recent growth and the latter a main trunk. But both those central elements of the Society's life had their roots solidly in the past. At times, one received more emphasis than the other, as successive Councils responded to larger movements in American society; and, despite the fact that the consultants denied reversing the principal recommendation of the goals conference, another of those shifts was now under way.

By the late 1970s, many within ASME felt themselves under siege. And it was not only industry engineers who were threatened by the power of federal regulatory agencies or consumer advocates. Codes and standards income funded a number of headquarters activities and the men responsible for the Society's budget began to think of contingency plans. Those were the loudest voices the management consultants heard and, at least in part, their recommendation that ASME make the technical sphere of action its central focus was a defensive strategy designed to protect the Society's tax status and its income.

On the eve of its one-hundredth anniversary, the American Society of Mechanical Engineers found itself in a situation not unlike that of a half-century earlier, when it celebrated its fiftieth birthday. Engineers in 1980 no longer enjoyed the widespread public confidence that the sophisticated weaponry of the Second World War had stimulated. They could no longer count on the status that had grown as America's prosperity swelled to luxurious proportions; and the public seemed already to have forgotten that remarkable affirmation of the nation's technological prowess when its spacemen walked on the moon. In a similar way, ASME's leaders in the 1930s also faced a crisis in technological confidence. But when Ralph Flanders determined to defend mechanization from the charge that its effects were uncivilizing and that it had caused the Great Depression, he was able to integrate the Society's social and technical purposes by describing the organization's role as part of a historical process. Holley and Thurston had seen it that way, too. ASME's founders called the Society to life to achieve two principal objectives and, as time would make manifest, they often conflicted with each other. Yet each has been an essential element and the way they have intertwined has framed the Society's existence.

Writing ASME's History

A POSTSCRIPT

For a good many years before its one-hundredth anniversary there had been an interest at ASME in a history of the organization that would follow up Frederick Remsen Hutton's 1915 account. The idea for his book had originated in 1903 and the plan was to have it available for the Society's twenty-fifth anniversary in 1905. Hutton's successor, Calvin Winsor Rice, who had a deep interest in museums of the history of technology, wrote a brief article in 1930 for the fiftieth anniversary issue of *Mechanical Engineering* entitled 'Fifty Years of the ASME,' but that still left open the prospect of a more detailed study.

George A. Stetson, editor of *Mechanical Engineering*, thought the sixtieth anniversary presented a suitable occasion for publishing a Society history, and in 1938 he drafted a memo suggesting the notion to fellow staff members Clarence E. Davies, Ernest Hartford, and Clifford B. Le Page. Stetson also pointed out that something should be done while those 'closest to the events' of Rice's long tenure as secretary were still alive. Clarence Davies liked the idea, but he told Stetson his own belief was that the job should be done by 'someone outside the immediate scope of staff operation.' The obvious candidate was Joseph Wickham Roe, professor of engineering at New York University, an active member of the Society, author of a historical study of British and American tool builders, who had also written the life of James Hartness in ASME's biography series. But Roe decided not to take on the task, as did L.P. Alford, Henry L. Gantt's biographer.

That left Stetson, and when he retired in 1956, almost twenty years later, he started gathering materials for the work. Known to many in the Society as 'Prof' for his scholarly temperament, Stetson spent the next two years combing Council minutes and other records and he filled out thousands of file cards with the name of every person, committee, and subject that was in any way related to the Society. It was an approach that would have daunted any but the most faithful and Stetson was only able to complete a hundred pages of the first draft of a manuscript before he died.

When O.B. Schier became secretary of the Society, he hoped Colonel Davies might use his retirement to write a history, and a History Committee chaired by Lou Rowley was formed in 1961 to help advance the project, with the expectation that a book would be available by ASME's centennial in 1980. However, Clarence Davies found it difficult to work on the history of the organization he knew so well. The first problem was that he was so often called upon to do other jobs such as manage the United Engineering Center. But another handicap proved just as troublesome; he kept getting absorbed by historical incidents that side-tracked him from the main purpose. 'It will take me a bit of time,' he wrote Rowley, 'to be developed into a historian.' Davies came back to the task in 1967 when he was reappointed to ASME's staff, and Schier started up the History Committee again, with Rowley still in charge. Some progress was made in identifying subjects and categories of source materials, but the project bogged down once more.

Although no one could recognize it explicitly, the great stumbling block was that a complete, documented record of the Society's people and its programs – the sort of thing most institutional histories aimed at – was impossible to write. Davies intuitively sensed part of the difficulty when he reacted against an 'official' history because so many of them 'did not seem of much value.' But even if the extraordinary amount of detail that George Stetson's card file hinted at could have been managed in a book of reasonable size, the source materials on which to ground the enterprise were not there. For want of space and a program to preserve its historical resources, ASME's housekeepers had long before disposed of its membership records – one of the few sources of information about the rank and file – as they had also discarded the correspondence of its staff, of its elected officers, and any letters that came in from the membership. The financial records are incomplete and so are those of local sections, technical divisions, and of the Society's education and research activities.

Miscellaneous as they are, and derived mostly from the last thirty years, the records that remain still comprise a substantial amount and they proved useful to me. As the footnotes reveal, however, this study depended mostly on ASME Council minutes, its periodical publications, the *Transactions* and *Mechanical Engineering*, and upon those papers of Society leaders in archives and libraries or still in private hands. ASME records, whether published or not, often mask the truth in controversial matters. Yet Council minutes and the Society's publications are extremely good sources of information, as well as nice indicators of the organization's styles and rhythms. The personal correspondence of active figures in the drama, however, gives a substance and dimension to history that no other source provides. I used the papers of Frederick Winslow Taylor and Morris L. Cooke, for instance, to learn some uncomfortable facts that were not in the official record of that era, just as the Flanders papers gave me an enhanced regard for the sensitivity of many mechanical engineers in the 1930s to difficult human problems.

The connection between candid, original documents and balanced history is a fact engineers need badly to recognize. They are prone to care about the judgment of history, but the historians who make the judgments are dependent upon the documents they see. Clarence Davies complained, for example, that the complete truth of Morris L. Cooke's relations with the Society was never known. But, ironically, Davies saved practically none of his own records, which might have more evenly illuminated that episode in ASME's history, while Cooke preserved every scrap of his papers.

As part of the Centennial observance, the various subdivisions of the Society were encouraged to write their individual histories. Those kind of topically limited efforts have the potential to serve much better the legitimate desire for 'works of record' that a centennial stimulates. But local sections, technical divisions, and the other units of ASME are even more disadvantaged than headquarters concerning the source materials of their history. Their officers, who come and go each year, imagine that New York keeps what they find unable to save for themselves. Almost inevitably, therefore, the chronological coverage of the histories they prepared for the Centennial was fixed by the memories of those who wrote them. That I nonetheless found many of them valuable is a testimony to the care with which many engineers tried to reconstruct their past.

The footnotes in this book have been mostly limited to the identification of the source of quoted material, since it did not seem appropriate to me to lumber them with the additional references suitable to a study aimed primarily at scholars. Similarly, it seems unnecessary to catalogue all those published works that bear on the history of engineering and its institutions. But I am happy to acknowledge my intellectual debts to the literature I found most immediately to the point. Three recent studies proved especially helpful in connecting ASME's history to related issues in America's past. Burton Bledstein's *The Culture of Professionalism* (New York: W.W. Norton 1976), gave me very useful insight into the ambitions that professionalism served and a better sense of the scale of that movement. The other two studies, although they approached the subject from substantially different perspectives, helped me to see ASME in an institutional context that to a considerable degree reflected industrial capitalism's drive for order and system. One was Alfred D. Chandler jr's *The Visible Hand* (Cambridge, Mass.: Harvard University Press 1977), and the other was David F. Noble's *America by Design* (New York: Alfred A. Knopf 1977).

I also drew heavily on two well established investigations into the institutional history of American engineering. Anyone interested in the history of mechanical engineering must read Monte A. Calvert *The Mechanical Engineer in America, 1830–1910* (Baltimore: Johns Hopkins Press 1967), which I did again, to my benefit. And Edwin T. Layton's book *The Revolt of the Engineers* (Cleveland: The Press of Case Western Reserve University 1971), remains the best analysis there is of engineering politics in America. Finally, there is a splendid though little-known essay by Eugene S. Ferguson that aided me a great deal. Entitled 'Sense of the Past: Historical Publications of the American Society of Mechanical Engineers,' it was written in 1974 when Ferguson was a member of the Society's History and Heritage Committee. Happily, it has been recently printed in Richard S. Hartenberg ed *National Historic Mechanical Engineering Landmarks* (New York: ASME 1979). Characteristically, it is full of good information and wisdom.

Council Members

1880–1980

The dates of service may not always be exact, as the Society has used different conventions of recording terms of office. For instance, a president taking office at the annual meeting in December 1905, and holding office for a year, would in some cases be listed as president for 1905–1906 and in others for 1906. It was also not always possible to determine the month when someone began or ended a term, especially in situations where Council appointed a person to fill an unexpired vacancy.

Some committee and board names changed over time, but those changes are not indicated in each pertinent entry. However, the most important ones are the following: Publications and Papers became Publications in 1923; Awards and Prizes became Awards in 1926, Honors and Awards in 1936, and the Board of Honors in 1966; Safety Codes became Safety in 1927; the Membership Committee became Admissions in 1936; Education and Professional Status became Education in 1959; Membership Board became Member Interest and Development in 1974; Professional Affairs became Professional and Public Affairs in 1975; and the Committee on Organization became the Committee on Planning and Organization in 1976.

Abbott, Robert E. Vice President Communications 1977

Abbott, William L. Manager 1908 to 1910, President 1926, Chairman Relations with the Colleges 1935

Abramson, H. Norman Vice President Basic Engineering 1974 to 1978

Alden, George I. Vice President 1891 to 1893

Alden, V.E. Chairman Professional Conduct 1943 to 1944

Alford, Leon P. Vice President 1921 to 1922, Vice Chairman Awards 1928, Chairman Awards and Prizes 1938

Allardt, Ernst W. (Regional) Vice President 1958 to 1959, Director (at large) 1962 to 1966

Allen, Charles M. Manager 1929 to 1931, Vice President 1932 to 1933

Allen, John R. Vice President 1920

Allen, Robert C. Chairman Technology (Board) 1959, Director (at large) 1961 to 1964

Allen, W.F. Chairman Finance Committee 1976

Amstuz, J.D. Chairman Finance Committee 1957

Angus, Robert W. Vice President 1925 to 1926

Armitage, Joseph B. Director (at large) 1948 to 1950

Armstrong, G.S. Chairman Professional Conduct 1945

Astley, Wayne C. Vice President Power 1969 to 1971

Atkinson, Kerr Chairman Technology (Board) 1960

Aurand, Henry S. (Regional) Vice President 1958 to 1959

Babcock, George H. Manager 1881 to 1884, President 1887

Bacon, R.H. Chairman Technology (Board) 1957

Bailey, Alex C. Manager 1933 to 1935, Chairman Research Committee 1934, Vice President 1936 to 1937, President 1945, Chairman Public Affairs (Board) 1948 to 1949

Bailey, E.G. Chairman Research Committee 1941, President 1948

Baker, A.L. Chairman Standardization Committee 1941

Baker, Charles W. Vice President 1910 to 1911

Baker, Robert A. (Regional) Vice President 1975 to 1977

Baker, W.S.G. Vice President 1887 to 1889

Baldwin, Stephen W. Manager 1887 to 1890, Vice President 1890 to 1892

Ball, Frank H. Manager 1888 to 1891, Vice President 1894 to 1896

Bancroft, J. Sellers Manager 1910, Vice President 1916 to 1917

Barber, Everett M. Director (at large) 1958 to 1964

Barker, Joseph W. Chairman Organization Committee 1955, President 1956

Barnard, Niles H. (Regional) Vice President 1962 to 1965

Barr, B.G. (Regional) Vice President 1966 to 1970

Barr, John H. Manager 1916 to 1918

Barrus, George H. Vice President 1905 to 1906

Basford, G.M. Manager 1907 to 1909

Bateman, George Frederick Chairman Publications and Papers 1939

Batt, William L. Chairman Meetings and Program 1930, Manager 1931 to 1933, Vice President 1934 to 1935, Chairman Awards

and Prizes 1935, President 1936, Chairman Public Affairs (Board) 1950

Bauer, Charles A. Manager 1894 to 1897

Bauer, H.J. Chairman Finance Committee 1956

Bausch, Carl L. Manager 1938 to 1940, Chairman Awards and Prizes 1944 to 1945

Bayles, A.L. Chairman Membership (Board) 1962

Bayles, James C. Secretary 1880

Beichley, F. Wendell (Regional) Vice President 1968 to 1970

Beitler, Samuel R. Chairman Constitution and By-Laws 1941 and 1943, Chairman Committee on Local Sections 1944, (Regional) Vice President 1946 to 1947

Belcher, W.E. Chairman Membership (Board) 1961

Belz, Ray E. Chairman Honors (Board) 1961 to 1963, Vice President Education 1968 to 1970

Benninghoven, Rhein (Regional) Vice President 1966

Benjamin, Charles H. Vice President 1917 to 1918

Bennett, Robert A. (Regional) Vice President 1971 to 1975, Treasurer 1976 to 1978

Bergman, Donald J. (Regional) Vice President 1961 to 1963

Bergman, Elmer D. Director (at large) 1955 to 1961, Chairman Codes & Standards (Board) 1960 to 1961, President 1964 to 1965

Berman, Irwin Vice President Communications 1976 to 1980

Billings, C.E. Vice President 1893 to 1895, President 1895

Black, Archibald Chairman Committee on Professional Divisions 1929

Black, Henry M. Vice President Education 1970 to 1972

Blackall, Frederick S., jr Director (at large) 1947 to 1950, Chairman Codes & Standards (Board) 1950, President 1953, Chairman Finance Committee 1962

Blake, A.D. Chairman Constitution and By-laws 1931

Bliss, Collins P. Chairman Standardization Committee 1927 to 1929, Chairman Meetings and Program 1934

Blizard, John Chairman Library Committee 1941 to 1945

Bond, George M. Manager 1888 to 1891, Vice President 1909 to 1910

Borden, Thomas J. Vice President 1888 to 1890

Boyer, Francis H. Manager 1899 to 1902

Bradley, Frank L. Chairman Publications and Papers 1942 to 1944, Director (at large) 1954 to 1957, Chairman Organization Committee 1958

Bradner, Mead Chairman Technology (Board) 1965, Chairman Constitution and By-laws 1970, Chairman Organization Committee 1971

Braisted, Paul (Regional) Vice President 1976 to 1978

Brashear, John A. Manager 1899 to 1902, President 1915

Breckenridge, Lester P. Vice President 1908 to 1909

Brigman, Bennett M. Manager 1935 to 1937, Vice President 1938

Brill, George M. Manager 1904 to 1907, Vice President 1911 to 1912

Brizzolara, R.D. Chairman Constitution and By-laws 1939

Bromley, Charles H. Chairman Constitution and By-laws 1926 to 1927

Brooks, H.W. Chairman Committee on Professional Divisions 1931

Brown, J. Calvin Manager 1944, (Regional) Vice President 1946 to 1949, President 1951

Browne, William H. Vice President General Engineering 1968, Vice President Research 1969 to 1971, Chairman Constitution and By-laws 1973 to 1974, Chairman Awards and Prizes 1975, Chairman Organization Committee 1976

Buckingham, E. Chairman Standardization Committee 1934

Burbank, Edward W. Chairman Relations with the Colleges 1936, Manager 1937 to 1939

Burchard, A.W. Chairman Finance Committee 1908

Bushnell, Fred N. Manager 1918 to 1920

Butler, H.W. Chairman Membership Committee 1932

Byrne, William H. (Regional) Vice President 1955 to 1958, President 1961 to 1962, Chairman Organization Committee 1967

Caldwell, A.J. Manager 1907 to 1909

Campbell, C.E. Chairman Constitution and By-laws 1958, Chairman Technology (Board) 1964

Carley, Charles T., jr (Regional) Vice President 1972 to 1974, Chairman Organization Committee 1975

Carlson, Harold C.R. Director (at large) 1955 to 1958

Carman, Edwin S. Chairman Committee on Local Sections 1920, President 1921

Carpenter, R.C. Vice President 1909 to 1910

Carr, J.D. Chairman Membership (Board) 1964

Carter, Wilber A. Chairman Committee on Professional Divisions 1943, Director (at large) 1947 to 1948

Case, E.H. Chairman Education (Board) 1962 to 1963

Chalmers, John B. Chairman Safety Codes Committee 1938

Chandler, Thompson (Regional) Vice President 1954 to 1955, Chairman Constitution and By-laws 1963

Chapman, C.M. Chairman Standardization Committee 1926

Chiarulli, Peter Vice President Membership 1974 to 1978

Chick, Alton C. Chairman Relations with the Colleges 1942, Manager 1944 to 1945, (Regional) Vice President 1946 to 1947

Christie, A.G. Manager 1923 to 1925, Chairman Publications and Papers 1924, Vice President 1926 to 1927, Chairman Professional Conduct 1932 to 1933, President 1939

Christie, James Vice President 1902 to 1904

Christy, William G. Manager 1942 to 1944

Church, E.F., jr Chairman Relations with the

Colleges 1928 and 1934, Chairman Library Committee 1947

Church, William Lee Manager 1884 to 1887

Cisler, Walker Lee Chairman Public Affairs (Board) 1954 to 1955, President 1960

Clark, A.J., jr (Regional) Vice President 1975 to 1978

Clark, Wallace Chairman Organization Committee 1947 to 1948

Clegg, R.I. Chairman Constitution and By-laws 1923 and 1925

Coes, H.V. Chairman Finance Committee 1925 to 1929, Vice President 1927 and 1933 to 1934, Manager 1930 to 1932, President 1943

Coggin, Frederick G. Manager 1886 to 1889

Cogswell, William B. Manager 1881 to 1882

Collett, S.D. Chairman Membership Committee 1919 and 1924 and 1929

Comstock, L.K. Chairman Membership Committee 1930

Condit, Kenneth H. Chairman Publications and Papers 1928, Chairman Awards and Prizes 1933, Chairman Committee on Professional Divisions 1936, Manager 1939, Vice President 1940 to 1941, Chairman Organization Committee 1949

Conlon, William T. Chairman Finance Committee 1937

Conta, Lewis C. (Regional) Vice President 1966 to 1968

Coogan, Charles H., jr (Regional) Vice President 1958 to 1962

Cooke, Harte Vice President 1938 to 1939, Chairman Awards and Prizes 1939, Chairman Committee on Professional Divisions 1940

Cooke, Morris L. Manager 1915

Cooley, M.E. Vice President 1901 to 1903, President 1919

Coonley, Howard Chairman Codes & Standards (Board) 1947 to 1949

Copeland, Charles W. Treasurer 1881 to 1883, Vice President 1884 to 1886

Corbett, Charles H. Manager 1901 to 1904

Cornell, Donald H. Chairman Technology (Board) 1962, Chairman Constitution and

By-laws 1968, Vice President Membership 1970 to 1974

Couch, A.B. Vice President 1883 to 1885

Couch, Harold Kennan (Regional) Vice President 1963 to 1965

Cowing, G.R. Chairman Education (Board) 1951 to 1954

Cox, Robert W. (Regional) Vice President 1965 to 1969

Coxe, Eckley B. Vice President 1880 to 1881, President 1893 to 1894

Cramp, E.S. Vice President 1896 to 1898

Crandall, Stephen H. Vice President Basic Engineering 1978

Crawford, David F. Manager 1911 to 1913

Crede, Charles E. (Regional) Vice President 1956 to 1958

Croft, Huber O. Chairman Relations with the Colleges 1940, Manager 1941 to 1943

Crotts, Marcus B. (Regional) Vice President 1967 to 1969

Cucullu, Lionel J. Director (at large) 1951 to 1954

Cunningham, James D. Chairman Committee on Local Sections 1928, Manager 1930 to 1932, Vice President 1933 to 1934, President 1950

Cunningham, Richard G. Vice President Education 1974 to 1976

Dale, O.G. Chairman Publications and Papers 1925

Daniels, F.H. Vice President 1902 to 1904

Daugherty, Robert L. Manager 1926 to 1928, Vice President 1929 to 1930

Davidson, Charles J. Manager 1912 to 1914

Davies, C.E. Secretary 1934 to 1957

Davis, Charles A. (Regional) Vice President 1965 to 1967

Davis, E.F.C. Vice President 1891 to 1893, President 1895

Davis, Harvey N. Manager 1929 to 1930, Vice President 1931 to 1932, Chairman Meetings and Program 1937, President 1938

Davis, J.H. Director (at large) 1955, Chairman Education (Board) 1957

Day, Emmett E. (Regional) Vice President 1962 to 1966

Dean, Francis Vice President 1895 to 1897

Deeds, E.A. Vice President 1922 to 1923

Degler, H.E. Chairman Relations with the Colleges 1944

Dehoff, Gerry B. (Regional) Vice President 1966 to 1968

Denton, James E. Manager 1889 to 1892

Derby, R.E. Chairman Finance Committee 1968

Dickie, Alexander J. Manager 1932 to 1934

Dickie, George W. Manager 1895 to 1898, Vice President 1915 to 1916

Diederichs, Herman Chairman Awards and Prizes 1936

Diemer, Hugo Chairman Professional Conduct 1939

Dillaway, Robert B. Vice President Industry 1970 to 1972

Dixon, Robert M. Chairman Finance Committee 1911 to 1918

Dixon, Walter F. Chairman Committee on Professional Divisions 1932

Dodge, James M. Manager 1891 to 1894, Vice President 1900 to 1902, President 1903

Dolan, Thomas J. (Regional) Vice President 1959 to 1961

Doolittle, Harold L. Manager 1931 to 1933, Vice President 1934 to 1935

Dorner, Frederick H. Manager 1928 to 1930, Vice President 1932 to 1933

Doty, Paul Manager 1927 to 1929, Chairman Committee on Local Sections 1929, Vice President 1930 to 1931, President 1934

Dow, Alex Vice President 1907 to 1908, President 1928

Draper, Eaton H. (Regional) Vice President 1961 to 1965

Dreyfus, E.D. Chairman Publications and Papers 1929

Drucker, Daniel C. Vice President Communications 1969 to 1971, President 1973 to 1974

Dryden, H.L. Chairman Technology (Board) 1952

Dudley, Samuel W. Chairman Meetings and Program 1929, Chairman Publications and Papers 1935 to 1936, Manager 1937 to 1939

Dudley, William Lyle Chairman Committee on Local Sections 1935 to 1936, Manager 1936 to 1938, Vice President 1939 to 1940

Duggan, Herbert G. (Regional) Vice President 1969 to 1971

Dupont, A.T. Chairman Constitution and By-Laws 1942

Durand, William F. Vice President 1912 to 1913, President 1925

Durfee, W.F. Manager 1883 to 1886, Vice President 1896 to 1898

Duzer, R.M. Van, jr Chairman Honors (Board) 1953

Earle, Samuel B. Manager 1938 to 1940, Vice President 1941 to 1942

Eastwood, E.O. Manager 1924 to 1926, Vice President 1927 to 1928

Eaton, Paul E. Manager 1941 to 1943, (Regional) Vice President 1948 to 1949, Director (at large) 1952 to 1953

Eckart, W.R. Vice President 1883 to 1885

Eckhardt, Carl J. (Regional) Vice President 1949 to 1952

Eggleston, Herbert L. Chairman Committee on Local Sections 1941, Manager 1942 to 1943

Ehbrecht, Adolf Chairman Membership (Board) 1959 to 1960

Eidmann, F.L. Chairman Awards and Prizes 1934

Elliott, Ben George (Regional) Vice President 1953 to 1956, Chairman Public Affairs (Board) 1957

Ellis, Daniel S. Manager 1945, Director (at large) 1946 to 1947

Elmer, William Manager 1926 to 1928, Vice President 1929 to 1930

Elsey, W.R. Chairman Research Committee 1945

Ely, S.B. Chairman Committee on Local Sections 1923

Ely, Theo N. Vice President 1881 to 1882

Emery, Charles E. Vice President 1881 to 1883

Emmons, Howard W. Vice President Basic Engineering 1966 to 1970

Engdahl, R.B. Chairman Technology (Board) 1963
Ennis, W.D. Vice Chairman Constitution and By-laws 1928, Chairman Constitution and By-laws 1929, Treasurer 1935 to 1944
Eschenbrenner, Gunther P. Vice President Industry 1978
Eshelman, Joseph W. Manager 1940 to 1942, Vice President 1943 to 1944
Fackler, Warren C. (Regional) Vice President 1979 to 1980
Faig, John T. Chairman Education and Training for Industry 1924 to 1929 and 1933
Falk, H.S. Chairman Education and Training for Industry 1934
Feiker, F.M. Chairman Meetings and Program 1932
Felton, Edgar C. Manager 1898 to 1901
Fernald, Robert H. Manager 1917 to 1919, Vice President 1920 to 1921, Chairman Power Test Committee 1936
Field, Crosby Chairman Committee on Professional Divisions 1937 to 1938
Finch, Rogers B. Secretary 1972 to 1978
Finlay, Walter S., jr Vice President 1923 to 1924
Fish, E.R. Manager 1924 to 1926, Vice President 1927 to 1928, Chairman Professional Conduct 1936 to 1937
Fisher, Earl V. (Regional) Vice President 1979 to 1980
Fisher, Elbert C. Manager 1920 to 1922
Fisher, George W. Manager 1881 to 1883
Flagg, Stanley G., jr Manager 1911 to 1913
Flanders, Ralph E. Chairman Publications and Papers 1926 to 1927, Manager 1927 to 1929, Vice President 1930 to 1931, President 1935
Fletcher, Andrew Manager 1890 to 1893
Folsom, Richard G. Director (at large) 1959 to 1962, President 1972 to 1973
Ford, Louis R. Chairman Membership Committee 1938
Forest, George M. Chairman Finance Committee 1920
Forstall, Walton (Regional) Vice President 1966 to 1970, Chairman Constitution and By-laws 1975 to 1978

Forsyth, Robert Manager 1891 to 1894
Forsyth, William Manager 1888 to 1891
Foster, Jack W. (Regional) Vice President 1973 to 1975
Foster, W.C. Chairman Public Affairs (Board) 1958 to 1959
Franck, Clarence C., sr Director (at large) 1960 to 1962
Fraser, David R. Vice President 1897
Freeman, Clarke Chairman Meetings and Program 1938, Manager 1939 to 1941, Vice President 1942 to 1943
Freeman, John R. Vice President 1902 to 1904, President 1905
French, T.E. Chairman Standardization Committee 1944
Freund, C.J. Chairman Education and Training for Industry 1936 to 1937
Fried, George Vice President General Engineering 1979 to 1980
Friederich, Allan G. (Regional) Vice President 1969 to 1971
Fritz, John Vice President 1882 to 1884, President 1896
Fulweiller, W.H. Chairman Research Committee 1932 to 1933
Funk, Nevin, E. Chairman Research Committee 1936 to 1937, Chairman Meetings and Program 1943, (Regional) Vice President 1947
Gagg, Rudolph F. Chairman Meetings and Program 1939, Vice President 1944 to 1946, Chairman Technology (Board) 1947
Gaither, Robert B. Vice President Education 1976 to 1978
Gantt, H.L. Manager 1909 to 1911, Vice President 1914 to 1915
Gates, P.W. Vice President 1907 to 1908
Gates, R.M. Chairman Meetings and Program 1927 to 1928, Manager 1929 to 1931, Vice President 1932 to 1933, President 1944
Geier, Fred A. Manager 1918 to 1919
Gillis, H.A. Manager 1901 to 1904
Gilmore, Quincy A. Vice President 1880
Goetzenberger, R.L. Chairman Education and Training for Industry 1945, Chairman Education (Board) 1947 to 1949, Director (at large) 1953 to 1956

Goff, J.A. Chairman Honors (Board) 1964

Goland, Martin Director (at large) 1963 to 1966, Vice President Communications 1966 to 1969

Goldschmidt, O.E. Chairman Membership Committee 1934

Gomff, A.M. Chairman Constitution and By-laws 1947, Chairman Membership (Board) 1949 to 1950

Goodale, A.M. Manager 1898 to 1901

Gordon, Alexander Vice President 1890 to 1892

Gorton, Charles E. Manager 1926 to 1928, Chairman Membership Committee 1927, Vice President 1929 to 1930

Goss, W.F.M. Manager 1900 to 1903, Vice President 1910 to 1911, President 1913

Graf, Samuel H. Manager 1944 to 1945, Director (at large) 1946, (Regional) Vice President 1950 to 1953

Graff, W.M. Chairman Safety Codes Committee 1935 to 1936

Granniss, E.R. Chairman Safety Codes Committee 1945

Graser, T.N. Chairman Constitution and By-laws 1964

Grasse, Harold (Regional) Vice President 1959 to 1962

Gratch, Serge Chairman Awards and Prizes 1968 to 1969, Vice President Research 1973 to 1977

Graves, Benjamin P. Director (at large) 1950 to 1953, Chairman Codes & Standards (Board) 1952 to 1953

Graves, Gilman L., jr (Regional) Vice President 1974 to 1976

Greene, Arthur M., jr Manager 1914 to 1916, Vice President 1917 to 1918, Vice Chairman Awards 1929, Chairman Awards and Prizes 1930 and 1932

Gregory, William B. Manager 1917 to 1919, Vice President 1921 and 1932 to 1933

Grinnell, Frederick Manager 1887 to 1890

Gunby, Frank M. (Regional) Vice President 1948 to 1951

Habach, George F. Director (at large) 1965 to 1966, Vice President General Engineering 1966 to 1969, President 1968 to 1969

Hahn, Gordon R. (Regional) Vice President 1959 to 1961

Haines, H.S. Manager 1896 to 1899

Hall, James A. Chairman Committee on Local Sections 1926, Manager 1934 to 1936

Hall, N.A. Chairman Education (Board) 1960 to 1961

Haney, Jiles W. Chairman Committee on Local Sections 1934, Manager 1935 to 1937

Hanhart, Ernest H. (Regional) Vice President 1952 to 1953, Chairman Organization Committee 1961 to 1962

Hanley, William A. Chairman Committee on Local Sections 1927, Manager 1928 to 1930, Vice President 1931 to 1932, Chairman Relations with the Colleges 1938, President 1941

Hanson, K.P. Chairman Constitution and By-laws 1951, Chairman Honors (Board) 1965 to 1966, Chairman Awards and Prizes 1966 to 1967

Happing, E.L. Chairman Honors (Board) 1950

Harder, John E. (Regional) Vice President 1974 to 1977

Harkins, H. Drake Chairman Honors (Board) 1958 and 1960

Harlow, James H. Chairman Membership (Board) 1953 to 1954, (Regional) Vice President 1963 to 1966, President 1966 to 1967

Harper, A.C. Chairman Education and Training for Industry 1943 to 1944

Harrington, E.W. Chairman Constitution and By-laws 1955

Harrington, John L. Chairman Constitution and By-laws 1920 to 1922, Vice President 1921 to 1922, President 1923

Hartness, James Manager 1910 to 1912, Vice President 1913, President 1914

Hartridge, A.L. Chairman Finance Committee 1964 to 1965

Hatch, T.F. Chairman Safety Codes Committee 1940 to 1941

Hawkins, George A. Director (at large) 1955 to 1958

Hawkins, John T. Manager 1886 to 1889

Heath, W.C. (Regional) Vice President 1959 to 1962

Heck, R.C.H. Chairman Awards and Prizes 1937

Helander, Linn Manager 1940 to 1942, (Regional) Vice President 1946 to 1948

Henning, Gus C. Manager 1896 to 1899

Herbert, L.E. Chairman Constitution and By-laws 1952

Herr, E.M. Vice President 1911 to 1912, Chairman Screw Threads Committee 1922 to 1923

Herreshoff, John B. Manager 1893 to 1896

Herron, James H. Chairman Committee on Local Sections 1922, Manager 1923 to 1925, Vice President 1935 to 1936, President 1937

Hersey, M.D. Chairman Research Committee 1943

Hess, Henry Manager 1912 to 1914, Vice President 1915 to 1916

Hewitt, William Manager 1884 to 1887

Higgins, M.P. Vice President 1901 to 1903

Hill, Hamilton A. Manager 1885 to 1888

Hill, W.H. Chairman Standardization Committee 1945

Hirshfeld, C.F. Manager 1930 to 1932, Vice President 1933 to 1934

Hoadley, John C. Manager 1880 to 1882

Hoagland, Frank O. Chairman Committee on Professional Divisions 1925, Vice President 1938

Hodgkinson, Francis Chairman Power Test Committee 1938 to 1944, Vice President 1940 to 1941

Holley, Alexander Lyman ASME Founder 1880, Vice President 1880 to 1882

Hollis, Ira N. Vice President 1912 to 1913, President 1917, Chairman Awards and Prizes 1924 to 1929

Hollman, M.L. Vice President 1894 to 1896 and 1903 to 1905, President 1908

Holloway, J.F. Manager 1880 to 1883, President 1885

Hopper, T.W. Chairman Constitution and By-laws 1965, Chairman Finance Committee 1970 to 1971

Hornbruch, F.W. Chairman Organization Committee 1965

House, Paul T., jr (Regional) Vice President 1974 to 1976

Houston, L.W. Chairman Finance Committee 1952

Hovey, O.E. Chairman Library Committee 1929

Howard, E.E. Chairman Constitution and By-Laws 1928

Howell, Glen H. (Regional) Vice President 1970 to 1972

Hulse, George E. Chairman Constitution and By-laws 1940, Manager 1941 to 1943

Humphreys, Alexander C. Manager 1908 to 1910, Vice President 1911, President 1912

Hunsaker, Jerome C. Vice President 1940 to 1941

Hunt, Charles Wallace Vice President 1892 to 1894, President 1898

Hunt, Robert W. Manager 1882 to 1885, President 1891

Hunter, John Manager 1914 to 1916, Vice President 1918 to 1919, Chairman Professional Conduct 1930 to 1931

Hunter, John A. Manager 1933 to 1935, Vice President 1936 to 1937

Hutchinson, Ely C. Manager 1929 to 1931, Vice President 1934 to 1935, Chairman Meetings and Program 1936, Chairman Honors (Board) 1949 and 1951

Hutton, Frederick R. Secretary 1883 to 1907, President 1907, Honorary Secretary 1908 to 1918

Iddles, Alfred Manager 1935 to 1937, Chairman Standardization Committee 1936, Vice President 1939 to 1940

Irwin, Kilshaw M. Chairman Finance Committee 1938 to 1939, Vice President 1940 to 1942, Chairman Organization Committee 1952

Jackson, D.C. Chairman Awards and Prizes 1943

Jackson, John Price Chairman Safety Codes Committee 1926 to 1928

Jackson, William B. Manager 1913 to 1915, Vice President 1916 to 1917

Jacobson, Eugene W. Director (at large) 1957 to 1960, Chairman Constitution and By-laws 1966 to 1967

Jacobus, David S. Manager 1900 to 1903, Vice President 1903 to 1905, President 1916

Jacobus, R.F. Chairman Membership Committee 1921

Jahncke, Ernest L. Vice President 1930 to 1931

Jappe, K.W. Chairman Finance Committee 1942 and 1944, Treasurer 1944 to 1949

Jarvis, Charles M. Vice President 1897 to 1899

Jensen, Allen H. (Regional) Vice President 1961 to 1965

Jeter, Sherwood F. Manager 1922 to 1924, Vice President 1925 to 1926

Johanson, Norman R. (Regional) Vice President 1972 to 1974

Johnson, H.H. Chairman Membership (Board) 1965

Johnson, J. Karl (Regional) Vice President 1979 to 1980

Johnston, Elmer (Regional) Vice President 1964 to 1965

Jones, Charles E. Chairman Education (Board) 1965, Vice President Education 1966 to 1968, Chairman Organization Committee 1972, Treasurer 1975

Jones, Edward C. Vice President 1920 to 1921

Jones, James E. (Regional) Vice President 1955 to 1956, Director (at large) 1961 to 1965

Jones, James R. Vice President Power 1979 to 1980

Jones, Washington Manager 1881, Vice President 1882

Judge, Thomas J. (Regional) Vice President 1962 to 1964

Judson, Harry H. Chairman Safety Codes Committee 1937

Kafer, John C. Manager 1894 to 1897, Vice President 1897 to 1899

Karelitz, G.E. Chairman Committee on Professional Divisions 1942

Karg, W.E. Chairman Membership (Board) 1957 to 1958

Kates, Edgar J. Chairman Publications and Papers 1945, Director (at large) 1946 to 1949, Treasurer 1957 to 1969

Katte, E.E. Manager 1911 to 1913, Vice President 1914, Chairman Committee on Professional Divisions 1922

Kavanaugh, W.H. Chairman Relations with the Colleges 1927, Chairman Constitution and By-laws 1938

Kearney, E.J. Chairman Standardization Committee 1930

Keenan, W.M. Chairman Library Committee 1933 to 1934

Keep, William J. Vice President 1903 to 1905

Keeth, J.A. Chairman Committee on Local Sections 1945, Director (at large) 1949 to 1952

Keller, E.E. Vice President 1914 to 1915

Kenerson, William H. Chairman Committee on Local Sections 1921, Vice President 1923 to 1924, Chairman Professional Conduct 1942

Kennedy, Julian Vice President 1916 to 1917

Kent, Robert T. Chairman Committee on Professional Divisions 1926 to 1928

Kent, William Manager 1885 to 1888, Vice President 1888 to 1890

Kenyon, Richard A. (Regional) Vice President 1978

Kerr, Arthur J. Chairman Committee on Local Sections 1940, Manager 1945, Director (at large) 1946 to 1947

Kerrerson, W.H. Chairman Relations with the Colleges 1922 to 1924

Kessler, Henry R. (Regional) Vice President 1951 to 1954

Kezios, S.P. Vice President Communications 1973 to 1976, President 1977 to 1978

Kimball, A.L. Chairman Meetings and Program 1942

Kimball, D.S. Chairman Meetings and Program 1919, Manager 1920 to 1921, President 1922

Kinderman, W.J. Chairman Constitution and By-laws 1962, Chairman Organization Committee 1963

Knight, G.L. Chairman Finance Committee 1943

Knight, Kenneth T. (Regional) Vice President 1975 to 1977

Knocke, L.T. Chairman Standardization Committee 1943

Knoop, T.M. Chairman Membership Committee 1942

Kopf, J.L. Chairman Finance Committee 1941, Treasurer 1949 to 1957

Kotnick, George (Regional) Vice President 1976 to 1978, Chairman Regional Affairs Committee 1978

Laidlaw, Walter Manager 1905 to 1908

Landis, James Noble Chairman Committee on Local Sections 1942, Director (at large) 1946 to 1948, President 1958

Landreth, Olin Vice President 1885 to 1886

Lardner, Henry A. Chairman Library Committee 1922 to 1927 and 1930, Chairman Membership Committee 1935

Larkin, David Vice President 1945, Director (at large) 1946

Larkin, F.V. Chairman Publications and Papers 1931, Chairman Relations with the Colleges 1939, Manager 1943

Larkin, W.H. Chairman Membership (Board) 1955 to 1956, Director (at large) 1960 to 1961

Latham, Allen, jr Vice President Research 1967 to 1969

Lauer, Conrad Chairman Meetings and Program 1926, Manager 1927 to 1929, Vice President 1930 to 1931, President 1932

Lawrence, John H. Chairman Committee on Professional Divisions 1923, Manager 1925 to 1927, Vice President 1928 to 1929, Chairman Finance Committee 1949

Lea, Robert E. Director (at large) 1954 to 1957

Leavitt, E.D., jr Vice President 1881, President 1883

Leland, Henry Martin Manager 1913 to 1915

Lewis, Orval L. (Regional) Vice President 1966 to 1968, Vice President Industry 1974 to 1978, President 1978

Lewis, Wilfred M. Vice President 1901 to 1903

Libey, Samuel H. Chairman Relations with the Colleges 1930, Chairman Membership Committee 1943

Lieb, John W., jr Manager 1903 to 1906, Vice President 1907 to 1908

Lindemann, Walter C. Manager 1936 to 1938

Ling, F.F. Vice President Research 1977 to 1980

Little, John W. (Regional) Vice President 1959 to 1961

Liversidge, Horace P. Manager 1922 to 1924

Lofts, David Vice Chairman Finance Committee 1930 to 1931, Chairman Finance Committee 1932

Long, C. Hardy (Regional) Vice President 1971 to 1973

Loring, Charles H. Vice President 1885 to 1887, President 1892

Louden, J. Keith Chairman Technology (Board) 1961, Director (at large) 1964 to 1966, Vice President Industry 1966 to 1968

Lovely, John E. Chairman Standardization Committee 1942, Vice President 1945, Director (at large) 1946

Low, Fred R. Vice President 1919 to 1920, Chairman Power Test Committee 1922 to 1924 and 1927 to 1935, President 1924

Luce, A.W. Chairman Safety Codes Committee 1942

Lyford, F.E. Chairman Membership Committee 1944, Chairman Finance Committee 1948

McAlpin, William J. (Regional) Vice President 1964 to 1966

McBride, Thomas C. Chairman Constitution and By-laws 1932 to 1933

McBryde, Warren H. Vice President 1938 to 1939, President 1940

McCullough, W.R. Chairman Finance Committee 1969

McEachron, K.B., jr Chairman Education (Board) 1958

McEwan, Thomas S. Manager 1942 to 1944, Vice President 1945 to 1946, (Regional) Vice President 1946 to 1948

McFarland, Walter M. Vice President 1905 to 1907

Machell, Arthur R., jr Vice President Codes & Standards 1972 to 1976

McKiernan, John W. (Regional) Vice President 1965 to 1973, Vice President Membership 1978

McKinney, R.C. Manager 1902 to 1905, Vice President 1905 to 1907

McLain, R.H. Chairman Membership Committee 1936 to 1937

McLean, William G. (Regional) Vice President 1954 to 1955, Chairman Constitution and By-laws 1966, Vice President Codes & Standards 1970 to 1972, Chairman Organization Committee 1973, Chairman Technical Affairs 1976 to 1977

McLeish, Duncan R. Vice President General Engineering 1975 to 1978

McMillan, L.B. Chairman Meetings and Program 1925, Manager 1928 to 1929

McMinn, Bryan T. (Regional) Vice President 1956 to 1957

McWilliams, Bayard T. (Regional) Vice President 1970 to 1972

Magruder, William T. Vice President 1926 to 1927

Main, Charles T. Manager 1915 to 1917, President 1918, Chairman Professional Conduct 1924 and 1928

Mangan, J.L. Vice President Power 1975 to 1978

Manning, Charles H. Manager 1892 to 1895, Vice President 1895 to 1897

Marburg, L.C. Chairman Relations with the Colleges 1926

Marlowe, Donald E. (Regional) Vice President 1960 to 1963, Chairman Public Affairs (Board) 1962 to 1965, Vice President Professional Affairs 1967 to 1969, President 1969 to 1970, Treasurer 1975

Martin, Harold E. Chairman Organization Committee 1950, Director (at large) 1951 to 1954, Chairman Finance Committee 1959

Martin, Kingsley L. Vice Chairman Finance Committee 1932, Chairman Finance Committee 1933 to 1934

Maruska, Gerald F. (Regional) Vice President 1971 to 1973

Mason, J.A. Chairman Constitution and By-laws 1969

Mattice, Asa M. Manager 1903 to 1906

Maxwell, Maxwell C. Chairman Relations with the Colleges 1931 to 1932

May, De Courcy Manager 1900 to 1903

Meadows, Roscoe, jr (Regional) Vice President 1965 to 1967

Mehringer, F.J. Chairman Awards and Prizes 1976

Meier, E.D. Manager 1895 to 1898, Vice President 1898 to 1900 and 1910, Chairman Finance Committee 1905 to 1907, President 1911

Meixner, H.H. Chairman Education (Board) 1955 to 1956

Melville, George W. Vice President 1895 to 1897, President 1899

Merchant, M.E. Vice President General Engineering 1973 to 1975

Merrick, J.V. Vice President 1883 to 1885

Metcalf, William Vice President 1882 to 1884

Miller, Alten S. Chairman Library Committee 1931 to 1932

Miller, Earle C. (Regional) Vice President 1966 to 1970, Chairman Organization Committee 1974, President 1976 to 1977

Miller, Frank W. Chairman Constitution and By-laws 1948, (Regional) Vice President 1956 to 1957, Chairman Finance Committee 1963

Miller, Fred J. Manager 1904 to 1907, Vice President 1908 to 1909, President 1920, Chairman Professional Conduct 1925

Miller, L.B. Manager 1893 to 1896

Miller, Ralph Chairman Constitution and By-laws 1959 to 1960, Chairman Organization Committee 1964, (Regional) Vice President 1968 to 1970

Miller, Spencer Manager 1915 to 1917, Vice President 1918 to 1919

Milligan, M.W. (Regional) Vice President 1977 to 1978

Mills, Robert W. (Regional) Vice President 1970 to 1974

Miskin, H.C. Chairman Finance Committee 1977

Moen, Walter E. (Regional) Vice President 1967 to 1969

Mole, H.E. Chairman Membership Committee 1941

Monroe, W.S. Chairman Standardization Committee 1933

Moody, Arthur M.G. Vice President Membership 1966 to 1968

Moody, Lewis F. Director (at large) 1947 to 1948

Moore, Lycurgus E. Treasurer 1880 to 1881, Secretary 1880

Moore, R.S. Manager 1901 to 1904

Morehouse, J. Stanley Chairman Honors (Board) 1956 to 1957

Morgan, Charles H. Manager 1884 to 1887, President 1900

Morgan, David W.R. Vice President 1944 to 1945, Chairman Membership (Board) 1951 to 1952, Director (at large) 1953 to 1954, President 1955

Morgan, Joseph, jr Vice President 1886 to 1888

Morgan, Thomas R., sr Manager 1886 to 1889

Morris, Henry G. Vice President 1887 to 1889

Morrow, L.C. Chairman Publications and Papers 1933 to 1934

Morton, Henry Vice President 1882 to 1884

Morton, Roscoe W. Manager 1943 to 1945

Moultrop, I.E. Manager 1909 to 1911, Vice President 1913 to 1914, Chairman Professional Conduct 1929

Mowery, H.W. Chairman Safety Codes Committee 1930 to 1933

Moxley, Stephen D. (Regional) Vice President 1951 to 1952

Mueller, W.C. Chairman Standardization Committee 1940

Mullen, T.Y. Chairman Professional Practice 1964

Muller, Edward A. Manager 1925 to 1927, Vice President 1928 to 1929

Muller, Henry N. (Regional) Vice President 1960 to 1962, President 1965 to 1966, Chairman Organization Committee 1970, Treasurer 1971 to 1974

Mumford, Albert R. Chairman Library Committee 1938 to 1940, (Regional) Vice President 1947 to 1950

Murphy, F.L. Chairman Awards and Prizes 1972

Muschamp, George M. Chairman Constitution and By-laws 1953, Chairman Organization Committee 1956 to 1957, Director (at large) 1962 to 1966

Nagler, Forrest (Regional) Vice President 1949 to 1950

Nason, Carleton W. Manager 1889 to 1892

Naylor, N.A., jr Chairman Professional Practice 1961

Neal, R.S. Chairman Constitution and By-laws 1934

Nee, Raymond M. Chairman Professional Practice 1965, Vice President Professional Affairs 1969 to 1973

Nelsen, Robert (Regional) Vice President 1962 to 1964

Nelson, E.W. Vice President Professional Affairs 1973 to 1975, Chairman Staff Benefits 1978

Nelson, J.W. Chairman Membership Committee 1922

Neuendorff, Richard (Regional) Vice President 1966

Newcomb, Charles L. Manager 1919 to 1921, Chairman Professional Conduct 1926, Vice President 1927 to 1928

Newman, C.L. Chairman Finance Committee 1975

Nimmer, Frederick W. (Regional) Vice President 1976 to 1978

Noble, Alfred Manager 1913 to 1914

Nordberg, B.V.E. Director (at large) 1946

Nordmeyer, L.C. Manager 1921 to 1923

Norris, Henry M. Manager 1921 to 1923

Noyes, Jonathan A. Vice President 1944 to 1945

Oberg, Erik Chairman Finance Committee 1921 to 1924, Treasurer 1925 to 1935, Chairman Meetings and Program 1940

O'Brien, Eugene W. Manager 1932 to 1933, Vice President 1935 to 1936, Chairman Relations with the Colleges 1941, President 1947, Chairman Honors (Board) 1955

Ohle, E.L. Chairman Constitution and By-laws 1924, Manager 1934 to 1936

Olive, T.R. Chairman Technology (Board) 1954

Orrok, George A. Manager 1912 to 1914, Chairman Publications and Papers 1919 to 1923

Ott, Dudley E. Vice President Power 1971 to 1975

Otte, Karl H. (Regional) Vice President 1967 to 1969

Page, Raymond J. (Regional) Vice President 1974 to 1976

Pankhurst, Jno. F. Vice President 1890 to 1892

Parker, James W. Chairman Meetings and Program 1933, Manager 1936 to 1938, Vice President 1939 to 1940, President 1942, Chairman Education (Board) 1950

Parker, John (Regional) Vice President 1974 to 1978

Parmakian, John Director (at large) 1962 to 1966

Parsell, R.L. Chairman Constitution and By-laws 1944 to 1945

Parsons, Harry De B. Manager 1916 to 1918

Partington, James Chairman Committee on Professional Divisions 1924

Pasini, Albert C. (Regional) Vice President 1950 to 1951 and 1956 to 1957, Director (at large) 1952 to 1955

Paulus, J. Donald Vice President Industry 1972 to 1974, Treasurer 1975 to 1976

Pavia, E.H. (Regional) Vice President 1973 to 1975

Pearson, Harry R. (Regional) Vice President 1953

Peck, Clair B. Chairman Committee on Professional Divisions 1934, Chairman Publications and Papers 1940 to 1941, Vice President 1942 to 1943, Chairman Technology (Board) 1948 to 1949

Peck, E.C. Chairman Standardization Committee 1922 to 1925

Penniman, Abbott L., jr Director (at large) 1948 to 1951

Penrose, Charles Chairman Committee on Local Sections 1925

Perkinson, T.F. Chairman Technology (Board) 1958

Perrin, Arthur M. Director (at large) 1959 to 1962, Treasurer 1969 to 1971

Peterson, Rudolph E. Director (at large) 1964 to 1966, Vice President Research 1966 to 1967

Peterson, Vernon A. (Regional) Vice President 1955

Pfisterer, G.E. Chairman Constitution and By-laws 1930

Phelps, Dudley F. (Regional) Vice President 1965 to 1967

Philbrick, H.S. Chairman Professional Conduct 1934

Pickering, Thomas R. Vice President 1892 to 1894

Piez, Charles President 1930

Pigott, R.J.S. Chairman Research Committee 1923 to 1928, Vice President 1937 to 1938, President 1952

Place, C.R. Chairman Membership Committee 1926

Plunkett, Charles T. Vice President 1917 to 1918

Plunkett, Robert Vice President Basic Engineering 1970 to 1974, Chairman Technical Affairs 1974 to 1975

Polk, Louis Chairman Codes & Standards (Board) 1954 to 1958, Director (at large) 1953 to 1959

Pope, John T. (Regional) Vice President 1978

Pope, Joseph Chairman Finance Committee 1954, Director (at large) 1955 to 1959, Chairman Public Affairs (Board) 1960 to 1961

Porter, Charles T. Manager 1882 to 1895, Vice President 1886 to 1888

Porter, Hollis P. Manager 1922 to 1924

Potter, Andrey A. Chairman Relations with the Colleges 1929, President 1933, Chairman Public Affairs (Board) 1952

Pratt, Francis A. Vice President 1880 to 1881

Prentice, D.B. Chairman Relations with the Colleges 1933, Chairman Committee on Local Sections 1939

Prescott, Frederick M. Manager 1905 to 1908

Prouty, Frank H. Manager 1938 to 1940, Vice President 1941 to 1942

Pryke, J.K. Chairman Professional Practice 1962

Purcell, Thomas E. Vice President 1943, (Regional) Vice President 1948 to 1949, Director (at large) 1950 to 1953

Pusey, C.W. Manager 1892 to 1895

Quinn, James D. Vice President General Engineering 1969 to 1972

Rae, Thomas W. Secretary 1881 to 1882

Rautenstrauch, Walter Chairman Research

Committee 1922, Chairman Finance Committee 1935 to 1936

Raynal, Alfred H. Manager 1899 to 1902

Reed, John C. (Regional) Vice President 1950 to 1951

Reeder, Harry C. (Regional) Vice President 1979 to 1980

Rees, Robert I. Chairman Meetings and Program 1935 to 1936

Reid, William T. Vice President Research 1971 to 1973

Reist, H.G. Manager 1910 to 1912, Vice President 1914 to 1915

Rettaliata, John T. (Regional) Vice President 1951 to 1952

Reynolds, B. Chairman Membership (Board) 1965

Reynolds, Edwin Vice President 1892 to 1894, President 1902

Rhodes, Allen F. Vice President Industry 1968 to 1970, President 1970 to 1971, Chairman Staff Personnel 1974

Rice, A.L. Vice President 1922

Rice, Calvin W. Secretary 1906 to 1934

Rice, Richard H. Manager 1904 to 1907

Richards, Charles B. Manager 1881 to 1882, Vice President 1888 to 1890

Richards, Charles Russ Manager 1919 to 1921

Richmond, George Manager 1897 to 1900

Ricketts, Edwin B. Vice President 1941 to 1942

Riker, A.L. Manager 1907 to 1909

Rivenbark, Edwin L. (Regional) Vice President 1968 to 1970

Robert, James M. Manager 1944 to 1945, Director (at large) 1946

Roberts, Arthur, jr (Regional) Vice President 1949 to 1950

Roberts, Montague H. Chairman Publications and Papers 1938

Roberts, Percival Vice President 1893 to 1895

Roberts, Robert E. (Regional) Vice President 1970 to 1974, Vice President Professional Affairs 1975 to 1977, Chairman Organization Committee 1977, Chairman Technical Affairs 1978

Robertson, Richard B. (Regional) Vice President 1969 to 1971, President 1974 to 1975

Robertson, Roy C. (Regional) Vice President 1961

Robinson, A. Wells Manager 1896 to 1899

Robinson, S.W. Manager 1881 to 1884

Rockwood, George I. Manager 1903 to 1906, Vice President 1924 to 1925

Roe, J.W. Chairman Meetings and Program 1924, Chairman Committee on Professional Divisions 1930, Chairman Awards and Prizes 1940 to 1941

Roe, Kenneth A. Chairman Finance Committee 1966 to 1967, (Regional) Vice President 1969 to 1971, President 1971 to 1972, Chairman Staff Benefits 1974 to 1977, Chairman Staff Personnel 1975 to 1976

Rosa, Ercole (Regional) Vice President 1979 to 1980

Rosenburg, Richard B. (Regional) Vice President 1972 to 1978

Roscoff, P.S. (Regional) Vice President 1977 to 1978

Rothermel, U. Amel (Regional) Vice President 1963 to 1965

Rowley, Louis N., jr Chairman Technology (Board) 1950, Chairman Finance Committee 1955, Director (at large) 1958 to 1961, Chairman Organization Committee 1966, President 1967 to 1968, Chairman Constitution and By-laws 1972

Roy, James S. (Regional) Vice President 1976 to 1978

Royer, D.L. Chairman Safety Codes Committee 1939

Russel, W.S. Manager 1893 to 1896, Vice President 1897 to 1899

Ryan, William F. Chairman Publications and Papers 1937, President 1957

Sackett, Robert L. Chairman Education and Training for Industry 1930 to 1932, Manager 1933 to 1935, Vice President 1936 to 1937, Chairman Membership Committee 1940

Sadwith, Howard M. Chairman Finance Committee 1978

Sague, James E. Vice President 1915 to 1916, Chairman Professional Conduct 1930

Samans, Walter Chairman Standardization Committee 1937 to 1938

Sams, James H. (Regional) Vice President 1957 to 1958

Sanco, W.J. Manager 1909 to 1911, Vice President 1924 to 1925

Sanders, Newell Manager 1902 to 1905, Vice President 1928 to 1929

Sanford, G.E. Chairman Safety Codes Committee 1929

Sargent, Henry E. Vice President 1919 to 1920

Saunier, W.P. Chairman Organization Committee 1951

Sawyer, W.H. Chairman Finance Committee 1947

Schaff, F.A. Chairman Finance Committee 1930 to 1931

Schier, O.B., II Secretary 1957 to 1972

Schwab, Charles M. President 1927

Schwanhausser, E.J. Chairman Finance Committee 1960 to 1961

Schwartz, F.L. Chairman Constitution and By-laws 1956, Chairman Organization Committee 1960

Scott, Earl F. Manager 1920 to 1922, Vice President 1923 to 1924, Chairman Professional Conduct 1938

Scott, F.W. (Regional) Vice President 1977 to 1978

Scott, Frank A. Manager 1924 to 1926

Scott, Irving M. Vice President 1891 to 1892

Sebald, Joseph F. Chairman Codes & Standards (Board) 1962 to 1964, Vice President Codes & Standards 1968 to 1970

See, Horace Vice President 1885 to 1887, President 1888

Seely, Warner Chairman Education and Training for Industry 1940, Chairman Honors (Board) 1952

Sellers, Coleman, jr Manager 1890 to 1893

Sellers, Morris Manager 1887 to 1890

Sengstaken, J.H. Chairman Committee on Professional Divisions 1945

Seward, H.L. Chairman Awards and Prizes 1931

Seyler, George A. Chairman Education and Training for Industry 1935

Sharp, H.R. Vice President Professional Affairs 1977 to 1978

Sharp, Joel Vice President 1889 to 1891

Shearer, William A., jr (Regional) Vice President 1972 to 1974

Sheehan, William M. Chairman Committee on Professional Divisions 1944, Director (at large) 1948 to 1951

Shock, William H. Vice President 1881 to 1882

Shoemaker, Guy T. Manager 1940 to 1942, Vice President 1943 to 1944

Shoudy, W.A. Chairman Publications and Papers 1930 to 1931, Chairman Committee on Professional Divisions 1935, Vice President 1936 to 1937

Shumaker, Clifford H. (Regional) Vice President 1954 to 1958, President 1962 to 1963

Sibley, Robert Vice President 1922 to 1923

Sillcox, Lewis K. Chairman Standardization Committee 1931 to 1932, Chairman Library Committee 1937, Chairman Committee on Professional Divisions 1939, Chairman Meetings and Program 1944, Director (at large) 1947, President 1954, Chairman Honors (Board) 1959

Simonsen, J.M. (Regional) Vice President 1973 to 1975

Sluys, W. Van der (Regional) Vice President 1973 to 1974

Small, H.J. Vice President 1893 to 1895

Smith, Horace S. Vice President 1886 to 1888

Smith, J.F. Downie Director (at large) 1956

Smith, Jesse M. Manager 1891 to 1894, Vice President 1894 to 1895 and 1899 to 1901, President 1909, Chairman Constitution and By-laws 1919

Smith, Leslie S. (Regional) Vice President 1978

Smith, Lester C. Chairman Constitution and By-laws 1950 and 1954 and 1957, Chairman Organization Committee 1959, Director (at large) 1963 to 1966, Vice President Professional Affairs 1966 to 1967, Chairman Organization Committee 1969

242 Council Members

Smith, Oberlin Manager 1883 to 1886, President 1890

Smith, Ronald B. Chairman Technology (Board) 1955, Director (at large) 1958 to 1961, President 1963 to 1964

Smith, V. Weaver Chairman Organization Committee 1954, Director (at large) 1957 to 1960

Snelling, Henry H. Chairman Constitution and By-laws 1935 to 1937 and 1949, Vice President 1939

Soderberg, C.R. Chairman Publications and Papers 1943

Soule, R.H. Manager 1898 to 1901

Sowden, P.T. Chairman Committee on Professional Divisions 1933

Space, C.C. (Regional) Vice President 1977 to 1978, Vice President Professional Affairs 1979 to 1980

Spencer, C.G. Chairman Professional Conduct 1935

Sperry, Elmer A. President 1929

Spicer, Clarence W. Chairman Standardization Committee 1935

Springer, E. Kent (Regional) Vice President 1970 to 1972

Sprong, S.D. Chairman Membership Committee 1945

Stanwood, James B. Manager 1897 to 1900

Staszesky, F.M. Chairman Finance Committee 1972 to 1974

Stearns, Thomas B. Vice President 1912 to 1913

Stephens, James O. Vice President Communications 1971 to 1973

Stetson, George R. Vice President 1898 to 1900

Stevens, John A. Manager 1916 to 1918, Vice President 1919 to 1920, Chairman Boiler Code Committee 1922 to 1923

Stevenson, A.R., jr Chairman Education and Training for Industry 1941 to 1942, Manager 1943 to 1945, (Regional) Vice President 1946

Stewart, J.P. Chairman Finance Committee 1966

Stiles, Norman C. Manager 1895 to 1898

Stirling, Allan Manager 1881 to 1884, Vice President 1885 to 1887

Stott, Henry G. Manager 1908 to 1912, Vice President 1913 to 1914

Stover, Rolland S. (Regional) Vice President 1957 to 1958, Director (at large) 1961 to 1965

Streeter, Robert L. Chairman Research Committee 1930

Strothman, L.E. Vice President 1922

Sullivan, G.L. Chairman Relations with the Colleges 1945

Suplee, Henry H. Manager 1897 to 1900

Sutherland, Robert L. (Regional) Vice President 1965 to 1967

Swann, J.J. Chairman Finance Committee 1940 to 1945

Swasey, Ambrose Vice President 1900 to 1902, President 1904

Sweet, John Edson ASME Founder 1880, Manager 1882, President 1884

Switzer, F.G. Chairman Meetings and Program 1945

Symons, W.E. Chairman Finance Committee 1919

Talbott, John A. (Regional) Vice President 1972 to 1974

Tallman, Frank G. Manager 1905 to 1908

Taylor, Frederick W. Vice President 1904 to 1905, President 1906

Taylor, H. Birchard Vice President 1924 to 1925

Taylor, Stevenson Vice President 1899 to 1901

Tenney, E.H. Chairman Professional Conduct 1941

Thacker, Milton B. (Regional) Vice President 1967 to 1968

Theiss, Ernest S. (Regional) Vice President 1952 to 1953

Thielscher, Herman G. Manager 1945

Thom, George B. (Regional) Vice President 1961 to 1963, Chairman Organization Committee 1968, Chairman Awards and Prizes 1974

Thomas, Carl C. Manager 1921 to 1923

Thomas, Frank A., jr (Regional) Vice President 1971 to 1973, Chairman Organization Committee 1978

Thomas, Percy H. Chairman Library Committee 1928

Thompson, J.G.H. (Regional) Vice President 1975 to 1977, Chairman Regional Affairs Committee 1976

Thompson, John Manager 1892 to 1895

Thompson, Joseph C. Vice President Membership 1968 to 1970

Thompson, P.W. Chairman Technology (Board) 1951

Thompson, Willis F. (Regional) Vice President 1952 to 1955

Thorson, A.W. Chairman Technology (Board) 1956

Thurston, Robert H. ASME Founder 1880, President 1880 to 1882

Todd, James M. Chairman Committee on Local Sections 1932 to 1933, Manager 1934 to 1936, Vice President 1937 to 1938, Chairman Honors (Board) 1948, President 1949

Toltz, Max Manager 1915 to 1917, Vice President 1918 to 1919

Towne, Henry R. Vice President 1884 to 1886, President 1889

Townsend, David Vice President 1899 to 1901

Trinks, W. Chairman Research Committee 1942

Trowbridge, Roy P. Director (at large) 1964 to 1966, Chairman Codes & Standards (Board) 1965, Vice President Codes & Standards 1966 to 1968 and 1976 to 1978

Trowbridge, William P. Vice President 1881 to 1883

Trueblood, Walter D., III (Regional) Vice President 1974 to 1976

Trump, Edward N. Vice President 1905 to 1907, Chairman Professional Conduct 1927

Turck, F.B. Chairman Organization Committee 1953, Chairman Finance Committee 1958

Tutt, Charles L., jr (Regional) Vice President 1964 to 1966, Chairman Constitution and By-laws 1971, President 1975 to 1976

Upp, J.W. Chairman Safety Codes Committee 1922 to 1925

Vachon, Reginald I. (Regional) Vice President 1978

Vandegrift, Erskine, jr (Regional) Vice President 1970 to 1972

Vauclain, S.M. Vice President 1904 to 1906

Vaughan, H.H. Vice President 1911 to 1912 and 1923

Velzy, Charles O. Vice President Industry 1978 to 1980

Verplanck, D.W. Chairman Education (Board) 1959

Vopat, W.A. Chairman Membership (Board) 1963

Wagstaff, R.R. Chairman Professional Practice 1963

Wahlenberg, W.J. Chairman Meetings and Program 1941

Waitt, Arthur M. Vice President 1900 to 1902, Chairman Finance Committee 1909 to 1910

Waldron, Frederick A. Chairman Membership Committee 1920 and 1925 and 1931

Wallace, L.W. Vice President 1938 to 1939, Chairman Research Committee 1940, Chairman Honors (Board) 1947

Walsh, J.L. Vice Chairman Finance Committee 1929

Walsh, T.A., jr Chairman Safety Codes Committee 1934

Walton, Edward H. (Regional) Vice President 1962 to 1966, Vice President Professional Affairs 1969

Walworth, Arthur C. Manager 1894 to 1897

Warner, Worcester R. Manager 1891 to 1893, President 1897

Warren, B.H. Vice President 1898 to 1900

Warren, Glenn B. Director (at large) 1956 to 1958, President 1959, Chairman Awards and Prizes 1970 to 1971

Warren, William J. (Regional) Vice President 1974 to 1976

Webber, Samuel S. Secretary 1880, Manager 1902 to 1905

Weber, Arthur W. (Regional) Vice President 1958 to 1959, Chairman Education (Board) 1964 to 1965

Weber, Erich C. (Regional) Vice President 1976 to 1978

Webster, D.T., jr Chairman Professional Practice 1960

Webster, Hosea Chairman Membership Committee 1923 and 1928 and 1933

Weeks, George W. Vice President 1889 to 1891

Weisberg, Herman Chairman Research Committee 1944

Wellman, Samuel T. Manager 1885 to 1888, Vice President 1896 to 1898, President 1901

Wells, Frank C. Manager 1919 to 1921

West, Arthur Vice President 1908 to 1909

West, E.H. Vice Chairman Finance Committee 1927 to 1928

Westcott, Harry R. Chairman Committee on Local Sections 1931, Manager 1932 to 1934, Vice President 1935 to 1937, Chairman Honors (Board) 1954

Westinghouse, H.H. Manager 1889 to 1892, Vice President 1904 to 1906

Wetzel, Theodore A. Chairman Constitution and By-laws 1961, (Regional) Vice President 1963 to 1965

Weymouth, C.R. Manager 1917 to 1919

Weymouth, Thom R. Vice President 1931 to 1932

Whitaker, H.E. Chairman Finance Committee 1950 to 1951

White, A.E. Chairman Research Committee 1929, Manager 1943 to 1945

Whiting, S.B. Manager 1881, Vice President 1882 to 1883

Whitlock, Elliott H. Manager 1914 to 1916, Vice President 1934 to 1935

Whittemore, Herbert L. Manager 1931 to 1933

Whyte, F.M. Vice President 1909 to 1910

Wichum, Victor Chairman Committee on Professional Divisions 1941

Wickenden, T.H. Chairman Membership Committee 1948

Wilberding, M.X. Chairman Public Affairs (Board) 1953

Wiley, William H. Treasurer 1883 to 1925

Wilkinson, Ford L., jr Chairman Committee on Local Sections 1943, Vice President 1944 to 1945

Wilkinson, Thomas L. Chairman Committee on Local Sections 1924 to 1930, Vice President 1925 to 1926

Williams, Edward E. (Regional) Vice President 1946 to 1948

Williams, S.C. Chairman Finance Committee 1953

Wilson, Donald R. (Regional) Vice President 1968 to 1969

Wilson, R.S. Chairman Awards and Prizes 1977 to 1978

Windle, A.E. Chairman Safety Codes Committee 1943

Winterround, William H. Chairman Publications and Papers 1932, Manager 1939 to 1941, Vice President 1941

Wohlenberg, Walter J. Vice President 1943 to 1944

Wolf, Robert B. Vice President 1921 to 1922

Wood, B.F. Chairman Professional Conduct 1940

Wood, De Volson Vice President 1889 to 1891

Woodbury, C.J.H. Manager 1882 to 1885, Vice President 1887 to 1889

Woolrich, Willis R. Chairman Committee on Local Sections 1937 to 1938, Manager 1939 to 1941, Vice President 1942 to 1943

Worden, E.P. Chairman Library Committee 1935 to 1936

Worley, Robert W. (Regional) Vice President 1962

Worthington, C.C. Manager 1883 to 1886

Worthington, Henry R. Vice President 1880

Wright, Paul Manager 1925 to 1927, Vice President 1928 to 1929

Wright, Roy V. Chairman Meetings and Program 1921 to 1923, Manager 1923 to 1925, Vice President 1926 to 1927, President 1931, Chairman Relations with the Colleges 1937, Chairman Awards and Prizes 1942

Yarnall, D. Robert Manager 1918 to 1920, Chairman Committee on Local Sections

1919, President 1946, Chairman Public Affairs (Board) 1951

Yellott, J.I. Chairman Relations with the Colleges 1943

Yopp, Paul R. (Regional) Vice President 1953 to 1954

Young, Dana Chairman Awards and Prizes 1973

Younger, John Chairman Education and Training for Industry 1938 to 1939

Zwiep, Donald N. Vice President Education 1972 to 1974, President 1979 to 1980

Index